Praise for *First Grade Arithmetic*

Vivian Jones-Schmidt has taken her decades of experience as a Waldorf educator and transformed it into the first of its kind, step-by-step, picture-by-picture, day-by-day procedures of how to create a Waldorf math curriculum in grade one. This carefully done guide gives clear instruction on how to build a clear and effective lesson for new learners in elementary school. A start, as described by Jones-Schmidt in this wonderful book, will give any first grader a rich, deep, complete lesson in grounded understanding of all that goes on with numbers and processes. Waldorf school teachers, Waldorf charter school teachers, homeschoolers and all in between, will benefit enormously from this mindful series of instructions for all who are responsible for introducing youngsters to math for the first time formally!

–PATRICE O'N. MAYNARD
Director of Publications and Development Waldorf Publications at the
Research Institute for Waldorf Education (former Waldorf class and music teacher)

This book is an invaluable resource for teachers at every level of experience. It addresses all aspects of teaching arithmetic in first grade: background, curriculum, methodology, artistic work, and practical activities.

The book covers the four major topics: form drawing; the qualities of numbers; the arithmetic operations; and expansion* of the operations. Each chapter contains a detailed guide to help teachers to plan, present and review the topics. Suggestions for written and artistic work are provided, accompanied by clear, simple illustrations. In addition, each of the chapters includes suggestions for singing, recitation, and movement, as well as a description of practical lessons to bring the subject to life.

Vivian Jones-Schmidt has done the Waldorf movement a great service by providing such a clear, comprehensive, useful guide to teaching arithmetic in first grade.

–ROBERTO TROSTLI
Roberto Trostli has been a class and high school teacher and has served as a member of the Pedagogical Section Council. His written works include *Physics the Waldorf Way; Chemistry the Waldorf Way; Home Surroundings the Waldorf Way; Thy Will Be Done: The Task of the College of Teachers*; and *For the Love of Language: Recitation the Waldorf Way — A Manual for Teachers*

First Grade Arithmetic

A Waldorf-Inspired Step-by-Step Curriculum

⇢ Trillium Math ⇠

Vivian Jones-Schmidt

First Grade Arithmetic: A Waldorf-Inspired Step-by-Step Curriculum
ISBN: 979-8-9990892-0-5 softcover
Library of Congress Control Number: 2025911295

Copyright ©2025 by Lotus & Ivy, LLC
All Rights Reserved.

No part of this book may be reproduced in any written,
electronic, recording, or photocopy form without the prior
written permission of the publisher, except for the inclusion
of brief quotations in a review.

Illustrations by Vivian Jones-Schmidt
Edited by Sarah Barrett and Heather Parrish
Book Design by Bob Schram, Bookends Design

Printed in the United States of America

Published by
LOTUS & IVY, LLC
Atlanta, GA
lotusandivyvirtualclasses@gmail.com

Appreciation

With unending gratitude, I dedicate this book to:

Russell Watson, whose math classes at Brookville High School ignited my love for math. Though dormant for many years, that love flared up again when I became a Waldorf teacher.

Harry Kretz, a masterful math teacher and musician, who could hold a class of children or adults in the palm of his hand.

Sarah and Heather, whose faith that I could do this work made it possible.

And of course, to my loving, supportive, and joyful family.

To all the students who accompanied me on the journey of arithmetic and always challenged me to be a better teacher.

Table of Contents

I. INTRODUCTION • ix

II. PREPARATION • xv
- Your "Classroom" • xv
- Materials • xvii

III. LESSON OVERVIEW • xix
- Sample Lesson • xix

IV. BLOCK SEQUENCE FOR MORNING LESSONS
- Lessons A1 - A12: Form Drawing • 1
- Lessons B1 - B12: The Quality of Numbers • 43
- Lessons C1 - C12: Introducing the Four Operations • 67
- Lessons D1 - D12: Extension of Operations • 97

V. BLOCK SEQUENCE FOR SKILLS CLASSES
- Skills Block One: 1 - 8: Modeling with Plasticine and Beeswx • 129
- Skills Block Two: 1 - 8: Solidifying the Understanding of Numbers • 147
- Skills Block Three: 1 - 8: Beginning to "See" with Hands and Fingers • 167
- Skills Block Four: 1 - 8: Review of the Basic Concepts • 185

VI. APPENDICES • 201

VII. ABOUT THE AUTHOR • 251

Introduction

IS THERE A NEED FOR YET ANOTHER BOOK ON TEACHING ARITHMETIC? Well, yes. In over 40 years of hands-on teaching experience and research, I've realized that too many adults tend to think of themselves as not liking or being good at arithmetic; and of course, this attitude makes its way into their teaching. I believe that using this book and this approach may serve to support the young child's natural inclination to enjoy arithmetic. It is also possible that it will help you, as an adult who may be uncomfortable with arithmetic, learn to enjoy it. Regardless, it is VERY important that you follow this rule: *Never indicate, in front of a child, that you don't like arithmetic, that you don't feel comfortable with it, or that math makes you anxious.* In fact, it would be best if you began each block by enthusiastically claiming that you love arithmetic! Children tend to model their reactions on those of their grown-ups, so if you follow this rule, you might just avoid passing your own discomfort on to your child. Jumping right in, what is the Waldorf approach to teaching arithmetic? Why is it different from the way arithmetic is usually taught? Does this approach prepare children for more complex mathematics?

Waldorf Education, from the Beginning

To understand the Waldorf approach, it might be best to start at the beginning. In 1919, Europe was recovering from the devastations of World War I, in which millions of people (both soldiers and civilians) had been killed or wounded, and untold thousands of farms, homes, and villages destroyed. In fact, human remains and ordinance from this war are still being discovered in Europe today. That spring of 1919, representatives from Allied countries were planning to meet in June to determine responsibility for the travesties of World War I and the consequences to be imposed.

On the other hand, some far-seeing thinkers were cautioning against a punitive approach to the instigators of the conflict.* They pointed out that, although the decisions had been made by German leaders, the consequences of those decisions would have to be borne by civilian populations who had suffered greatly both during and after the conflict. One of these far-seeing thinkers was a man named Rudolf Steiner, who traveled throughout Europe speaking publicly of his own thoughts: that this was a time which cried out for social renewal, not punishment.

After one such lecture, Dr. Steiner was approached by the industrialist Emil Molt, the director of the Waldorf-Astoria Cigarette Factory in Stuttgart, Germany. Herr Molt asked Dr. Steiner if he had recommendations for an approach to education that would help people avoid armed conflict in the future. Dr. Steiner's response was a resounding, "yes."

In addition to continuing his rounds of public speaking, over the next few months Dr. Steiner created an approach to educating children based on an understanding of human beings as possessing the capacities of body, soul, and spirit. An articulation of this conviction, as well as the curriculum which flowed from it, was ready for teachers to work with by August of that year. The first "Waldorf" school opened its doors for the children of the workers in Emil Molt's factory in September of 1919.

World War I began in 1914 when Archduke Ferdinand of Austria was assassinated by a Serbian dissident. Because of the complex system of alliances at the time, Britain, France, Russia, and Belgium felt obligated to respond when Austria-Hungary declared war on Serbia. Germany supported Austria-Hungary.

Teaching Arithmetic in the Waldorf Curriculum

Arithmetic, of course, was an integral part of the school's curriculum. Dr. Steiner was totally focused on the importance of arithmetic and its connection to the moral life:

"Arithmetic and moral principles are two things that, in terms of logic, seem very removed from each other. It is not common to connect arithmetic with moral principles, because the logical connection is not obvious. It is obvious, however, to those who do not view the matter in terms of logic but in a living way that children who are introduced to arithmetic correctly will have a very different feeling of moral responsibility than those who were not..."

—THE SPIRITUAL GROUND OF EDUCATION, PP 70-73

As can be implied from this statement, the Waldorf approach stresses the importance of "economy" in teaching. In other words, we try to teach multiple concepts in one lesson. If we want our children to grow up to be morally responsible adults, let us always be conscious of this goal no matter what we are teaching. As you go through this book, you will discover many lessons in which an underlying morality is present.

How is the Waldorf approach to teaching arithmetic different from the way arithmetic is usually taught? Being aware of the moral content of arithmetical operations is certainly one way Waldorf arithmetic lessons differ.

Waldorf lessons are also set apart in terms of what aspects of arithmetic are taught and when they are taught. The approach that seems to live most strongly in other systems today is called the "spiral" approach. Children, from kindergarten on, are exposed to lessons in addition and subtraction, statistics, economics, measurement, time, money, place value, common fractions and decimal fractions. This philosophy of teaching comes from the work of Jerome Bruner in the 1960's. Bruner posited that, rather than stay on one topic until mastery is achieved, children would learn better if they were exposed to multiple skills every year, and those skills were reviewed and enlarged upon in each successive year. Hence the "spiral" metaphor.

In the Waldorf school, children spend Grades 1-3 learning about whole numbers and how they behave in all four arithmetical operations [addition, subtraction, multiplication, and division]. The work naturally becomes more complex each year as they are taught place value, regrouping and "long" multiplication and division. By fourth grade, they have developed a firm understanding of these operations using whole numbers and are well-prepared to understand the behavior of parts of numbers: common fractions.

In a way, Bruner's "spiral" is utilized, because every year follows a pattern: review previously introduced subject matter in the fall, and introduce new subject matter in the spring.

Overview of the Waldorf Arithmetic Curriculum

Throughout the grade school [1-8], all previously taught material is reviewed in the arithmetic skills classes as well as in Morning Exercises. Waldorf teachers understand that children need to return to foundational concepts regularly in order to truly incorporate them into their understanding. So we do not expect a child to understand concepts shortly after they are introduced. Rather, we look at proficiency as developing over time. We introduce the four operations in First Grade, and review and expand upon them in Second Grade. By the end of Third Grade, we expect children to be able to add, subtract, multiply and divide with ease and flexibility. But again, all of this work, for three years, is with whole numbers. Only when children are absolutely solid in their understanding of how whole numbers interact with one another are they introduced to common fractions in the Fourth Grade.

The Waldorf curriculum in mathematics grows with the child. Measurement, time, and money are taught by the end of Third Grade. Geometry is a constant theme from first grade on, transforming from freehand geometric drawing in the early grades to the study of classical theorems in the Middle School. Decimal fractions are introduced in the Fifth Grade, Business Math [in a sense, the introduction to Algebra] in Sixth Grade, and Algebra in Seventh and Eighth Grades. Each year's curriculum includes review and expansion of what was taught in previous years.

You might have noticed that following the developmental curriculum of the Waldorf school does not always align with any particular state educational standards. It is accurate to say that our curriculum exceeds as well as falls short of such standards in every grade.

However, by the end of elementary school, our students have a deep understanding of number concepts and how to work with them. Moreover, Waldorf students tend to approach new studies in mathematics with confidence and even enthusiasm, and their competency is on a par with or exceeds that of their chronological peers from other educational systems. Because we teach arithmetic as a subject that leads to one discovery after another, children come to expect a certain degree of adventure in their journey through mathematics.

Arithmetic and the Arts

Because this is a Waldorf curriculum, the arts are integrated into our teaching methods. Throughout the grades, lessons are taught in story form; and every story has a picture. Most morning lessons set aside a few minutes for drawing aspects of the lesson. Sometimes, the teacher will provide a model; sometimes, the students will draw their own impressions. The teacher pays careful attention to the children's work and offers suggestions and instructions for improvement, with the type of comment depending on the developmental stage of the child. In First Grade, for example, a child who typically draws only with pencil or ink might be instructed to use at least three colors.

Form Drawing classes from First Grade on explore two-dimensional geometry. Not only do the children learn to draw various forms, they also create them in movement—because we realize that geometry lives in the body. These forms are initially created as simple line drawings, which are beautiful in themselves. When color is added, they become beautiful in an entirely different way. Which colors complement one another, and which colors clash? Which color combinations do you like or dislike?

As the years progress, the forms become more complex. Celtic knots in Fourth Grade are a challenge for child and adult as well. From First through Fifth Grades, drawing is primarily free-hand without the use of instruments. Toward the end of Fifth Grade or at the beginning of Sixth Grade, formal geometric drawing with straight edge and compass is introduced. In the Sixth Grade Medieval History Block, for example, children might design their own rose windows. Geometry culminates in Eighth Grade with the construction of three-dimensional forms from two-dimensional "nets."

Musical cues can reinforce the multiplication tables/skip counting. Sing the "fours" to "Mary Had a Little Lamb," for example. You can make the numbers fit the tune! Then find tunes for the other tables. Mathematics is about patterns, and patterns adapt themselves to movement. "Twinkle, twinkle, little star" becomes "1…2…**3**, 4…5…**6**, etc." Learning about common fractions in Fourth Grade is a perfect time to learn how to read music: a whole note can be two half notes or four quarter notes.

A Guide as a Cookbook

If you cook, you probably have a collection of cookbooks. Recipes can be approached in at least two ways: you can follow the steps and use the ingredients exactly as printed, or you can adjust the recipe to suit the ingredients you feel will be better for your family. In a similar vein, this material is presented as a ***guide*** for teaching Arithmetic to first grade children. Lessons are intended to be suggestions, not commands.

Your work with this material will serve you and your child best if you always study each lesson before teaching it. In fact, it would be best to read through an entire "block" of lessons a week before teaching it, and then to study each lesson carefully the evening before presenting it. If you follow this approach, you'll be prepared in terms of materials, the words you'll use, what you'll draw on the chalkboard, etc. You will know, for example, if there's something in the lesson that will be too challenging or too easy for your child; and you'll be able to adjust the lesson accordingly.

Whether you are teaching in a Waldorf school or home-schooling your own child, you've chosen a rich and fulfilling vocation. It is my hope that using this guide will assist you in feeling confident in your teaching of arithmetic and able to relax and enjoy the both children and the subject.

–Vivian Jones-Schmidt

Preparation

TO STATE THE OBVIOUS: teaching requires preparation, whether we're teaching one student or thirty, children or adults. The success of a lesson will depend on how well the teacher is prepared as well as the "fit" between the teacher/teaching style and the student/s. Take some time to study this manual from start to finish so you'll have an overview of the program. Here are a few pointers that we hope will make your lesson go smoothly and feel successful.

Be sure you have all the materials ready for the lesson you plan to teach. This sounds simple and like a "no-brainer," but it will take effort.

Your child will need a physical space that will support their efforts to learn. This means finding or creating a child-sized desk and a chair that your child can sit in comfortably with their feet firmly settled on the floor. Create a "classroom" space in your home—a quiet space in which you and your child can work without interruptions or distractions. Check that the lighting is at a comfortable level—the more natural light, the better. This space does not need to look like a traditional classroom, but should allow for the necessary components to be met: quiet, good lighting, and child-sized furniture.

Materials

THE FOLLOWING MATERIALS will be used throughout the program. If an asterisk is present, this means that sources for the material and/or instructions for making your own are in the appendix.

Materials for Every Lesson:
- We recommend working in commercially available, 11" x 14" [landscape view] Main Lesson books with 48 pages and spiral binding. Loose pages are really hard to keep up with– or to keep. Number the pages, front and back, from 1 - 48, before the lessons start. Odd numbers will be on the fronts and even numbers will be on the backs. These are the numbers that will be referred to in this guide, so it will be easier for you to keep track if your pages are numbered like this.
- a 17" x 23" chalkboard, a set of colored chalk, and an eraser
- 2" beanbag [no larger, please]*
- copper rod, 24"-30" [These are usually available at a hardware store. Buy two "cane tips" to cover the ends for safety.]
- 9" play ball
- drawing mat*

Form Drawing:
- spiral ML book #1 [You'll use the same book for Skills Class work.]
- block and stick crayons

Quality of Number:
- spiral ML book #2 [You'll use the same book for the two "operations" blocks.]
- block and stick crayons
- drawing mat*

Introduction of Operations:
- spiral ML book [#2 as above]

- block and stick crayons
- drawing mat
- 12 counting stones or beans, etc. It's a good idea to keep these in a little bag or box.
- number cards, 1-12*
- subitizing dot cards, 1-8, in various combinations*
- two standard dice [1-6 dots]
- dominoes

Expansion of Operations:
- spiral ML book #2 [same as before]
- block and stick crayons
- drawing mat
- 24 counting stones or beans, etc., with container
- number cards, 1-100*
- subitizing dot cards, 1-12, in various combinations*
- bead rope*
- 4 standard dice
- dominoes
- "number path"*

Skills Classes:
- Painting supplies:
  - 11"x14" watercolor painting paper, 130-140 lb. weight
  - liquid watercolor paints, set of 6: yellow, gold, vermillion, carmine, cobalt, ultramarine
  - painting surface [a vinyl mat or a masonite board, larger than the painting paper]
  - two sponges and a spray bottle
  - a 1" flat brush
- Beeswax, 1"x4" sheets, set of 15
- Plasticine, set of 8 colors [Crayola makes a 2-lb. set of 8 colors.]
- block and stick crayons
- set of counting stones
- spiral ML book #1 [the same book you used for Form Drawing]

Sample Lesson: Lesson 4A

Lesson Focus: For the convenience of the adult, each lesson is numbered and categorized by subject/block. For example, the fourth lesson in Form Drawing will be labeled "A4." This title is followed by the focus of the lesson. We recommend starting with the Morning Exercises and progressing right into the content of the lesson rather than saying something like, "Today we're going to talk about the number 7." The children will thus experience the flow of the lesson rather than intellectualizing it.

Morning Exercises: Each class starts with the adult and the child engaging together in movement, speech, and/or singing. Each activity is listed in order and the accompanying movement is described. Research has shown the strong connection between movement and learning, especially in children.

Each set of exercises is repeated at the beginning of the lessons for a period of time, generally for at least two weeks. While we list specific exercises, as you observe your child you will see the exercises that your child really likes, those that your child finds challenging, and those that your child finds not challenging enough. Feel free to repeat an exercise your child likes; to put off for later those they find challenging; and to discard those they find not challenging enough. Keep all exercises handy, as you will bring back early exercises later in the year.

Lesson Content: In Waldorf classrooms, the teacher establishes a learning rhythm that extends over 2-3 days. For example, at the end of Day 1, the teacher presents the lesson [in story form] for the next day. Day 2 begins with a review/retelling of the story and continues with a guided drawing based on the story. On Day 3, we present a brief review to start the lesson, followed by written work with the lesson content. As you read through the lessons, you will see this alternation in process, even though our lessons are designed for a 3-day instead of a 5-day week.

As foundations for lessons, we have included stories as originally written, and from various sources. Waldorf teachers carefully study the stories beforehand so that they can tell the story rather than read it in class. This is so important. When you read a story, your focus and your eyes are on the words on the page. When you tell a story, you are engaging the child with eye contact, observing the child's response to the story and perhaps adjusting your words accordingly. This makes telling the story more of an interaction between you and the child.

Perhaps, as you read a story beforehand, you might feel that a particular image or activity might be disturbing to your child. In the past and in other cultures, childhood was very difficult; also, many traditional stories were meant to be told to adults, not to children. So references to wine, for example, or terrible injuries, might appear in the original versions. Feel free to adjust a story for your child if you feel uncomfortable with an aspect of it.

As you study each story, please do not feel that you must memorize it. Rather, create mental images of the story's events. Then, when you tell the story, "call up" the images in the order in which you created them. Using this technique truly enhances one's ability to learn a story.

Beginning and Ending the Lesson: Establish a pattern for beginning and ending each lesson. Before the first day of school, you will have collected all of the materials needed and placed them in a basket, a drawer, or on a shelf that the child can reach. This way, at the beginning and ending of the lesson, it will be easy for the child to stack their materials and place them in the storage area provided. To begin the lesson, make sure all materials are close at hand. Your child will stand behind their chair for the opening verse or song and then move into an open space for the morning exercises. At the end of the lesson, close your books, put your crayons/pencils back in their containers. After the supplies are put away, the child pushes in their chair and stands behind the chair for the ending song.

MAIN LESSON BLOCK A

Form Drawing

Form Drawing was introduced as a school subject by Rudolf Steiner in his early lectures for teachers. He in fact presented the first lesson in First Grade as a practice with the straight line and the curved line – the foundational forms for all alphabetical and numerical writing. The form of anything and everything in the world is a combination of straight and rounded lines, and children experience this on the first day of their Waldorf school experience.

There are various categories of forms, and many of them correspond to particular historical periods and cultures. Teaching a form relies on several steps, which are outlined in the first Form Drawing lesson [page 3]. Children [and adults] who practice form drawing develop an inner sense of symmetry and balance; their powers of observation are strengthened, as well as their understanding of how living things change as they grow. The ability to observe objectively is foundational for work with any of the sciences as well as the humanities. We depend on clear observation as we work to understand the natural world in biology, chemistry, and physics, as well as the human world in sociology, civics, and economics. As Van James, renowned Waldorf art teacher, has pointed out, form drawing serves children throughout their educational journeys and indeed, throughout their lives as it stresses balance, uprightness, and beauty.

The First Grade begins with Form Drawing because it is the perfect way to introduce children to the fundamental practices that are expected of them in school. For example, they are expected to hold a pencil in a particular way, to write from left to right, to stand up straight, to sit with feet on the floor, to follow the adult's directions, to work with concentration and without talking as they complete their assignments. [See Appendix for Information on Proper Pencil Grip]. We recognize that you might resist, for example, teaching your child to draw a straight vertical line–after all, haven't they been doing that since they were two? But following the instructions just described elevates drawing a straight line from an almost unconscious action to an intentional and challenging one. Moving from unconscious actions to conscious ones is an important task in any child's development.

Here we are underlining the support that form drawing gives to healthy brain development. Intentional activity, to mention only one such support, is critical for the maturation of executive function. In looking at the model form and working to match the movement of their hand with the movement of their eye, the child demonstrates the development of *intention* out of *interest*. If a child is interested in an activity or a subject, the child's work will be intentional rather than careless or indifferent. The teacher, working out of *enthusiasm*, models *interest* for the child.

As was indicated in the introduction, studying arithmetic is also related to moral development. In the Form Drawing block in particular, you will find many stories. These are to be told because they provide a background for drawing specific forms, but the stories also ask us to think about right and wrong, civil and uncivil behavior, cleverness and kindness. The story of the *Race Between Turtle and Bear,* for example, invites much discussion: What do you think of Bear's behavior toward Turtle? Turtle's response to Bear's behavior was clever, but was it the right thing to do? What do you think Turtle should have done?

As often as possible, the stories we reprint in this guide are the original stories as told in the original cultures. Read the stories all the way through during your lesson preparation. If an element of a story feels uncomfortable for you, change it or choose another story that illustrates the concept you'll be teaching. Not every story meets every child or family; and although we've tried to choose traditional stories that any family can use, we encourage you to be thoughtful in your choices and presentations.

You will find an emphasis on working with moving forms throughout the Waldorf curriculum. In addition to form drawing, for example, many of the morning exercises are movement exercises. As elucidated by Jeff Tunkey in his excellent book, *Educating for Balance and Resilience,* mathematics is inner movement. Success in arithmetic depends upon the ability to move numbers around and, in a sense, to imaginatively move oneself around numbers. We stress movement with children because movement is the foundation for understanding the operations of arithmetic. Form Drawing is much more than simply drawing a line on paper. The more you incorporate full body movement into your preparation for drawing, the more deeply the form will imprint into your child's body. This activity will not only support a healthy brain development, it will also lay the foundations for handwriting, free-hand drawing, and understanding geometric forms.

Please make sure you refer to these instructions whenever you teach a form to your child, whether it is in the Main Lesson or in a Skills Class.

Steps for drawing forms:
1. Draw the form on the teacher's chalkboard.
2. If possible, "make" the form with your body.
3. "Move" the form by walking it on the floor.
4. Draw the form with your arm, in front of your body, using the full range of motion. This is a large motor movement.
5. Practice drawing the form on the chalkboard.
6. Draw the form three times in the air over the page.
7. Draw the form on the page.
8. From time to time, you might add to the sequence: "drawing" the form on the child's back; asking the child to draw the form on the chalkboard; drawing the form in sand; creating the form with plasticine or beeswax. You can also "draw" the form in the air before showing it to the child on the chalkboard; ask the child to imagine what it will look like; then draw it on the chalkboard and ask the child if that form is the one they imagined.

Materials for the Form Drawing Block:
- 2" beanbag
- 9" play ball
- Drawing mat
- Spiral ML Book #1
- Block and stick crayons
- 17" x 23" chalkboard, chalk, and eraser

LESSON A1

Focus: Introduction to school and classroom routine. Tell the first story, ***"Beyond the Garden Wall"***. Illustration for the first form drawing: straight line and curved. [You'll draw the illustration today, and tomorrow, you'll draw the forms.]

Materials:
- ML book #1 for adult and child.
- Adult has finished the drawing on a 12" x 18" piece of drawing paper before class. Drawing mat for child.
- Easel for adult's book [The "easel" can be a music stand, or anything that will serve to display the adult's book so that the child can see it and the adult can easily refer to it.]
- Stick and block crayons.

Morning Exercises:

Through the Night [song]
Through the night, the Angels kept
Watch beside me while I slept.
Now the dark has gone away.
We are glad for this new day.

Morning Has Come [song]
Morning has come, night is away.
We rise with the sun and welcome the day.

Round and Round [song with movement]
Round and round, the Earth is turning,,
Turning always round to morning,
And from darkness round to light.

[Standing straight and tall, turn slowly in place until you are facing forward at the end of the song.]

Tall Trees in the Forest [verse with movement]
[Repeat three times, varying speed of movement once the child knows the verse.]

Tall trees in the forest;	*[Standing straight and tall, lift up on toes and reach high with your arms.]*
Pine cones on the ground.	*[Slowly bend over and "touch" the pine cones.]*
Tall trees in the forest,	*[Standing straight and tall.]*
They bend, they bend, they bend.	*[Bend over slowly and touch the floor.]*

Humpty Dumpty [verse with movement]
[Walk slowly forward, stepping on the accented syllables. Then reverse and walk backwards.]
Humpty Dumpty sat on a **wall**.
Humpty Dumpty had a great **fall**.
All the King's **hor**ses, and **all** the King's **men**
Could **not** put **Hump**ty to**gether** a**gain**.

Straight as a Spear [verse with movement]

Straight as a spear, I stand.	*[Standing as tall and straight as possible.]*
Strength fills my legs and arms.	*[Spread legs apart, stretch arms straight out to sides.]*
Warmth fills my heart with love.	*[Cross arms over heart.]*

[Swinging your arms around to each side, jump back into upright position, arms at sides.]

Stepping Stones [verse with movement]
[Step forward on accented syllables. Notice that some steps are slow, some faster.]
Step on **step**ping stones, **one, two, three**.
Step on **step**ping stones, **fol-low me**.
The **river** is **fast**, the **river** is **wide**.
We'll **step** on **step**ping stones **to** the **other side**.

I Am a Tall Tree [verse with movement]

I am a tall tree, reaching up to the sky.	*[Standing straight, stretch arms up.]*
When the wind blows, I lean with a sigh.	
I lean to the front,	*[Legs straight, lean to the front.]*
I lean to the back,	*[Legs straight, lean to the back.]*
I stand up straight without a crack.	*[Jump back into uprightness.]*

Pease Porridge Hot [verse with finger movement]
>[*Touch thumb to fingertips in turn on accented syllables, first with right hand, then with left, then both together. Start with thumb to index finger; continue to little finger, then go back one fingertip at a time to the index finger. Repeat. At the end of the verse, you will have touched each fingertip with your thumb twice.]]*

Pease porridge **hot, pease por**ridge **cold,**
Pease porridge in the **pot, nine days old.**

The Sun with Loving Light [verse with no movement]
>[*Stand up straight and tall, crossing arms over the heart, with reverence.]*

The Sun with loving light makes bright for me each day.
The Soul with Spirit power gives strength unto my limbs.
In sunlight shining clear, I do revere, O God,
The strength of humankind which you so graciously
Have planted in my soul,
That I with all my might
May love to work and learn.
From you stream light and strength.
To you rise love and thanks.

I Look at My Hands [verse with gestures]

I look at my hands with my fingers fine,	*[Hold hands out in front of you.]*
And I want to be proud that they are mine.	
For deep in my heart lies a golden chest	*[Cross hands over heart.]*
With secret treasures that no one can guess	
Unless my hands do their very best.	*[Hold hands out in front of you.]*

Warm Our Hearts [song to end Morning Lesson]

Warm our hearts, O Sun, and give	*[Cross hands over heart.]*
Light that we may daily live,	
Growing as we want to be:	
True and good and strong and free.	

Lesson Content:

1. The child sits down but the adult remains standing.

2. The adult tells the story, "Beyond the Garden Wall."

3. When the story is finished, say, "Now we're going to draw a picture from the story. Here is the picture." Show your completed illustration and say, "Here is the garden wall; here is the rose bush; here is the path." Give the child a drawing mat, their copy of ML book #1, and crayons.

4. Open your book to page 2, and make sure your child's book is also opened to page 2. [Pages should be numbered, in the center of the page at the bottom edge, before the first day of school. The first page of each book will be the title page, which will be completed in later lessons.]

5. Prepare to draw the illustration again, this time going slowly and narrating your work as you go. Say, "First, we're going to draw Mother Earth. We all stand on Mother Earth." With the "papa bear" [widest] side of the green block crayon, slowly draw along the bottom edge of the paper from left to right. Then go over the green with the brown block crayon. Explain, "Colors like to have friends. For Mother Earth, green and brown are friends."

6. With the blue stick crayon, draw the slats of the picket fence on the left side of the page. With the green block crayon, lightly draw the rose bush on the right side of the page; lightly go over the green with the yellow block crayon. Add the red roses with the red stick crayon.

7. With the black stick crayon, lightly add stepping stones from the end of the fence to the rose bush. Go over the black with the blue crayon. Now add blades of grass along the fence edge and between the stepping stones. With the gold block crayon, add a bold sun, with rays extending outward, in the middle of the top edge of the paper. With the blue block crayon, add swirls of blue for the sky, explaining that the air is always moving.

8. Close your books and put the materials away. Child stands and pushes in chair, standing behind chair. Sing "Warm Our Hearts" to end the morning lesson.

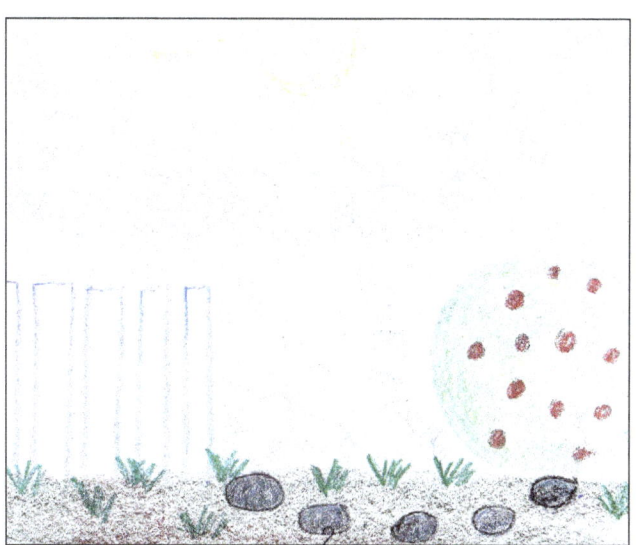
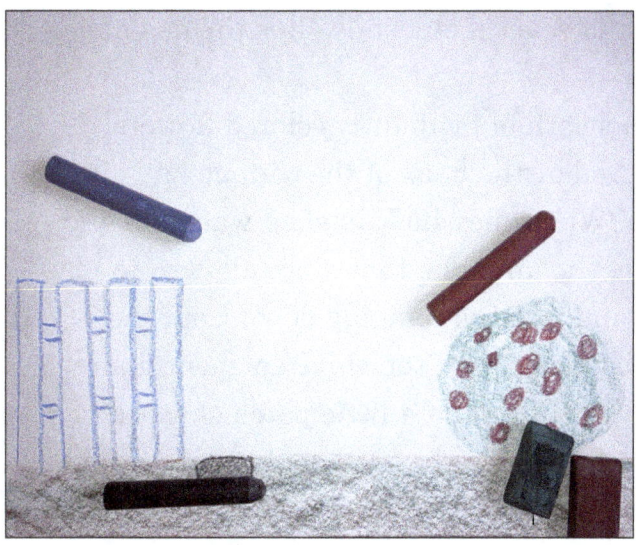

Beyond the Garden Wall

Lesson A1 and A2

By Vivian Jones-Schmidt

Once upon a time there was a small cottage nestled up close to a great forest. In this little cottage there lived a family: the parents and two children. All day long, every day, this family worked together. On one side of the cottage was their vegetable garden, where they grew lettuce, carrots, beets, kale, and sweet potatoes. On the other side of the cottage, they kept their animals inside a fence. They had a donkey, two pigs, and several chickens, plus the dog and cat that sometimes came inside the house.

The front of the cottage was bordered by a swath of brilliantly colored flowers: red, orange, yellow, white, purple, and pink. And at the back of the cottage, a white fence surrounded a play area for the children. When they had finished watering and weeding the garden, and watering and feeding the animals, they were allowed to play here inside the fence. Here there was a slide and a swing hanging down from the branch of a tree. There were large stones and stumps, and the children spent many happy hours jumping from one to the other. They even had a little patch of soft dirt that they could pour water into, and when they had mixed a nice mud, they could make mud cakes and little houses.

One day, one of the children happened to look over the fence. "Oh look!" shouted the child. "Where did that beautiful rose bush come from?" The rose bush was indeed beautiful, with shiny green leaves and bright, deep red roses. Neither of the children had ever seen it before. "We should pick a few of the roses for our parents," they said to one another. "If we aren't greedy, the rose bush won't mind if we take just a few."

Quietly and carefully, they opened the gate. There in front of them was a straight path leading to the rose bush. They danced along the path to the rose bush, but before they picked even one rose, they walked all around the bush to see how many roses it bore and which ones they would pick. Of course, roses have thorns, so they knew they would have to be very careful when they picked them. When they had picked only a few, they took them into their little cottage and put them in a vase of water for their table. And weren't their parents surprised when they saw this flash of color in their little home!

Focus: Review story. Draw the first forms: the straight line and the curved.

Materials:
- ML book #1 for child and for adult
- Easel for adult's book and drawing mat for child
- Stick crayons; block crayons
- Yellow chalk, chalkboard, and eraser.

Morning Exercises: Repeat in the same order as Day 1.

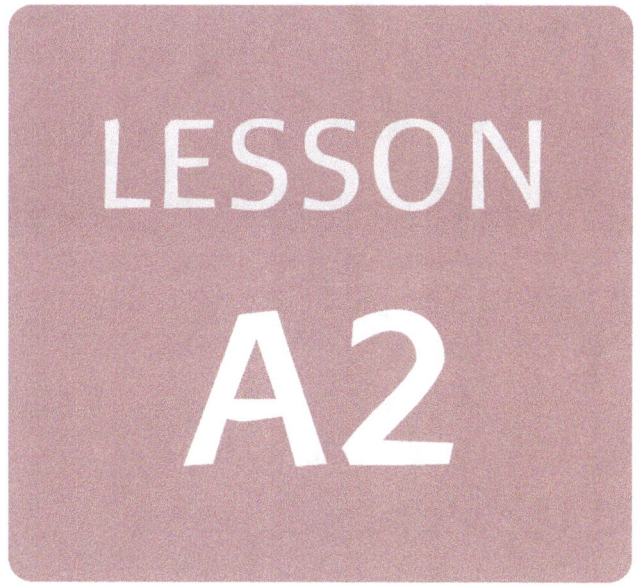

Lesson Content:

1 Story review. The review of the story should not be word for word, but rather "picture by picture." Begin with the child sitting up straight, feet on the floor. You can start with the first sentence of the story as you remember it. Then ask such questions as, "Where did this family live?" "How did the children help their parents?" "What was their favorite place?" "Why do you think the garden had a wall?" Continue for about 5 minutes total, until you get to the ladder and the path.

2 Draw the straight line and the curved line on your chalkboard, saying, "I call this line a straight line and this line a curved line." Ask the child to repeat the labels and then to stand. Ask the child to walk in a straight line and then a curved line.

3 Discuss where the child sees straight lines and curved lines in the room.

4 Take out ML book #1 and turn to **page 3**. With the "mama bear" [middle] side of the blue crayon, slowly draw a border on your page as the child observes. Draw a straight line, starting on the left edge of the paper, a line from " stars to stones." Repeat on the right edge of the paper. Now draw a line across the top edge of the paper, from "stars to stars," and across the bottom edge of the paper, from "stones to stones."

5 Now "draw" a straight line in the air in front of you, saying, "I'm drawing a straight line in the air." Then ask the child to "draw" the straight line in the air with you. In your imagination, divide page 3 in half. Now draw the straight line in the air three times above the page; then draw it on the left half of page 3 with the blue stick crayon.

6 Next, "draw" a curved line in the air in front of you, saying, "I'm drawing a curved line in the air." *This is a little tricky, as you want to draw the line in the direction*

 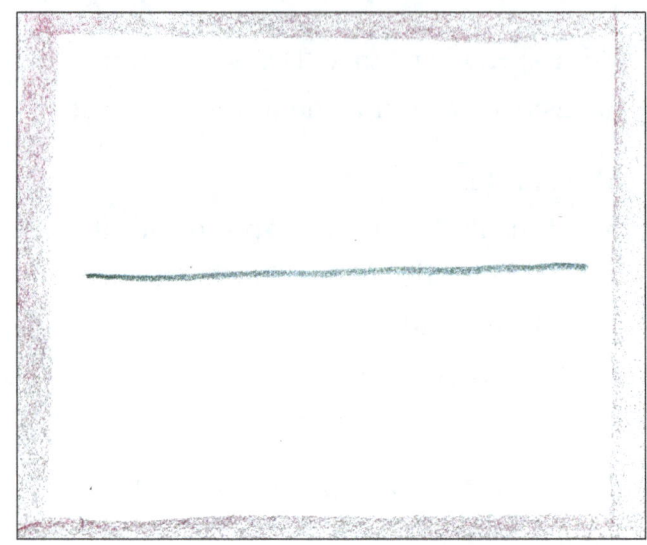

it will be drawn on the page–from the child's point of view. The line will be drawn on the page as: (. Stand next to, or in front of, the child so that you are using the same arm as the child uses to "draw" the form in the air.

7 Ask the child to "draw" the curved line in the air with you. Now draw the curved line three times in the air above the page before drawing it on the right half of page 3 with the red crayon. Your finished page will illustrate the two lines as: | (.

8 Say: "Now we've drawn a straight line and a curved line. Watch as I draw the next line." Then slowly and carefully draw a horizontal line on the chalkboard, from left to right. Say: "This is a straight line lying down." Ask the child to copy this line with their body, and ask how that feels. The child might respond that they feel like they're resting; and the line may then be called a "resting" line.

9 Prepare **page 4** in ML book #1 as above [4]. Say to the child, "Now I'm going to draw a resting line in my book." Draw the line three times in the air over the page, then in your book with a green stick crayon. Then ask the child to draw the line three times in the air over the page, then in their book with a green stick crayon.

10 Ask the child to close their book, put away their crayons in the box, stand up straight and push in the chair. Sing the song to end ML: "Warm Our Hearts."

First Grade Arithmetic—Teacher's Guide   13

Focus: Continue to explore the possibilities of straight line forms.

Materials:
- ML Book #1 for adult and child
- Easel for adult and drawing mat for child
- Stick and block crayons
- Yellow chalk, chalkboard, and eraser

Morning Exercises: Repeat in the same order as Day 1.

Lesson Content:

1. Child is seated. Open book to **page 3**. Pointing to each line in turn, say, "What do we call this line?"

2. Turn to **page 4**. "What do we call this line?"

3. Keep the books open. "Now we'll look at some new lines. Let me see you stand as straight as you can." [Child stands.] "Here is the line you make when you stand straight and tall." [Adult draws straight vertical line on the chalkboard.]

4. "You are standing on the Earth. We all stand on the Earth. This is how I draw you standing straight and tall on the Earth." [Adult draws straight horizontal line on the chalkboard, under the vertical line so that the vertical line is centered on the horizontal line.]

5. Child stands and "draws" the form in the air as in Lesson A2.

6. "Now we're going to draw these lines in our books." Using a block crayon on the "mama bear" side, draw lines on the edges of **page 5** "from stars to stones," "stars to stars," and "stones to stones."

7. Draw the vertical line in the middle of the page; draw the horizontal line under it.

8. "Now stand straight and tall. Imagine that you have a friend on each side of you, and reach out your arms to hug your friends." [This will look like a "plus" mark.] "This is how I draw you reaching out to your friends." [Draw the two lines, crossing in the middle, on the chalkboard.]

9. Follow the procedure as outlined above.

10. Open your books to **page 6**. Prepare the borders of the page. Draw the two lines in the middle of the page as the child follows along.

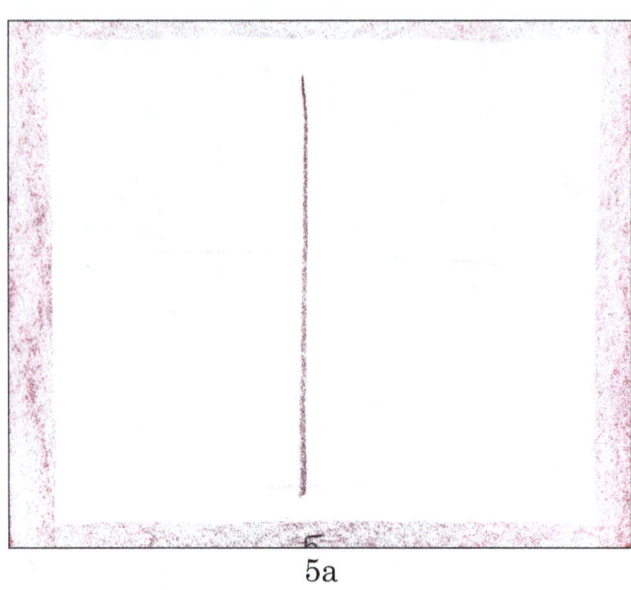

5a

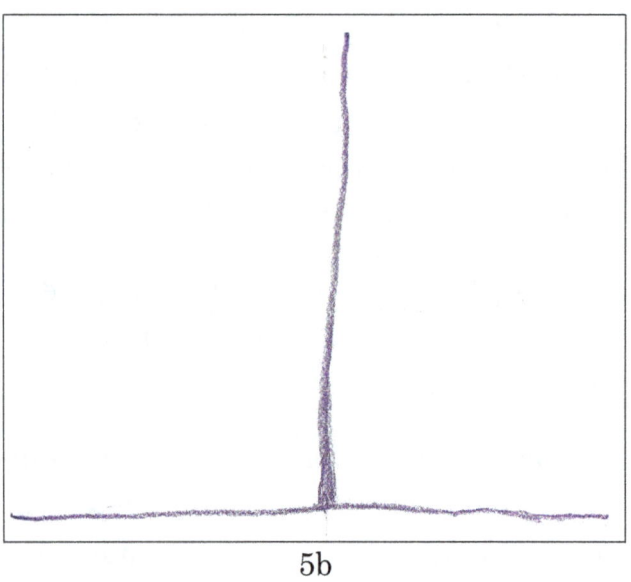

5b

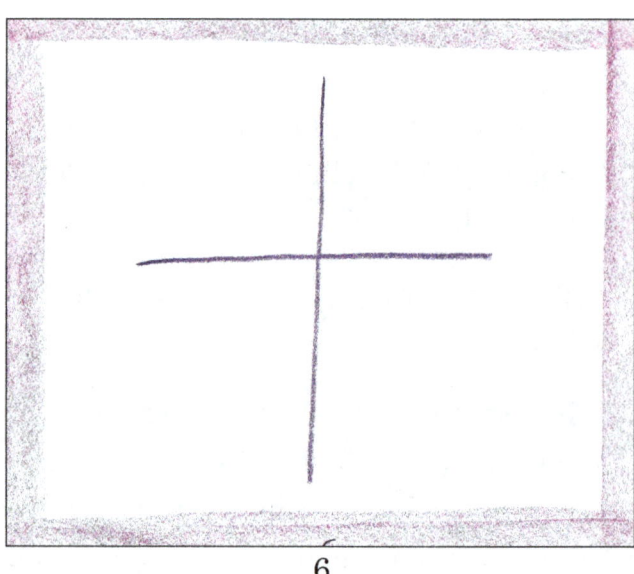

6

11 Close the books and put the materials away.

12 The child is sitting straight, feet on the floor, head up. The adult, standing, tells *"Finding the Money Tree."*

13 Stand together and sing, "Warm our hearts."

Finding the Money Tree

Lesson A3 and A4

Story in Public Domain

ONCE UPON A TIME in ancient China, there lived a family with two sons. The older son was lazy and avoided work whenever he could, while the younger son was always eager to help with even the hardest tasks. Nevertheless, the older son was the favorite son. When the parents died, they divided the land between their two boys, but they gave the best land to the lazy son. The younger son was left with land on a far hillside, rocky and dusty.

Now the older son continued as he had begun. He did not like to work, and he thought that the good things in life were meant to come to him without any effort on his part. So when the roof developed a tiny leak, he did nothing to repair it; and over time, the tiny leak became a stream of water coming into his house. When he did not plant any vegetables, no vegetables grew. When he did not repair the fence of the pigsty, the pigs ran away. So there he was, with water pouring into his house and nothing to eat. Even worse, he had no idea why these things were happening!

One day he visited his younger brother on the far away hillside. To his amazement, his brother's land was no longer rocky and dusty. He was living in a neat little stone cottage, with flowers all around and a vegetable garden to one side. His pigs were contentedly grunting in the pigsty down the hill; while the hill itself was flourishing with blossoming peach trees.

"Brother!" he cried. "How came you to have such good fortune? Do you have a money tree hiding in the orchard?"

"Oh yes," laughed his brother. "I have a very generous money tree!"

"I knew it," the older brother thought. "I'll come back at night and dig up that money tree and take it to my land. I deserve that fortune! After all, I am the older son."

A few nights later, the older brother took his shovel and hiked to the younger brother's farm. There he wandered through the orchard until he found the tree that he felt sure was the money tree. He dug it up and carried it back to his own land, where he planted it in good, rich soil. He tended the money tree as if it were a child, watering it often and telling it how beautiful it was. But no matter how tenderly he cared for the tree, in time, it only produced peaches. No money ever fell from its branches. So back to his younger brother he hiked.

"Brother!" he called. "You did not tell me the truth! Your money tree will not give me any money!"

"Oh my brother!" replied the younger brother. "Did you think that I really have a tree that showers me with money? Let me show you my 'money tree.'"

And the younger brother held out his hands and arms. "You see? Here are the branches of my "money tree." Pointing to his back and legs, he said, "Here is the trunk of my money tree. The trunk holds me up while my branches grab my tools and work in the soil. You have a money tree just like I do. You just need to use it!"

The older brother walked home slowly and thought about everything his younger brother had said. And very soon, he gathered tools and put his own "money tree" to work.

LESSON A4

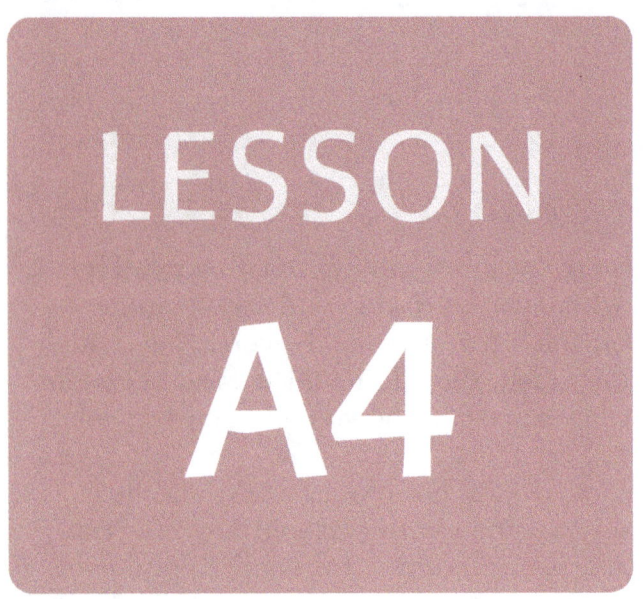

Focus: Continue to explore the possibilities of straight line forms.

Materials: The same materials as in Lesson A3.

Morning Exercises: Repeat in the same order as on Day 1.

Lesson Content:

1. Review the story, *"Finding the Money Tree."* [See Day 2 for ideas on how to proceed with a story review.]

2. Open your books to **page 5**. "What do we think of when we see this line?"

3. Turn to **page 6**. "What can we imagine when we see this line?"

4. "From our story, we have another imagination for straight lines. We stand on the Earth, like our drawing on **page 5**. But we stand with the Sky above us. This is how we draw a human being standing on the earth with the sky above.

5. The adult draws the form on the chalkboard. The child stands and "draws" the form in the air as above.

6. "Now we are going to draw this form in our books. Open your book to **page 7**." Prepare the page as before.

7. Adult carefully and slowly draws the form in the middle of the page. Child follows.

8. Close books and put materials away.

9. Child sits up straight, feet on floor, head up. Adult tells, *"The Runaway Rice Cakes."*

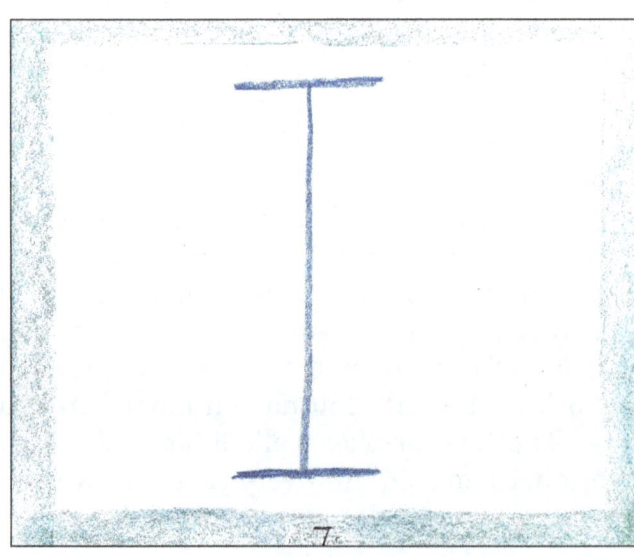

The Old Woman and the Runaway Rice Cakes

Lesson A4 and A5

Story in Public Domain

ONCE UPON A TIME in Old Japan, there lived a little old woman who made the best rice cakes in her village. Her little house was on the top of a hill, and people would come from all over the valley to ask for even a crumb of one of her rice cakes. She was happy to share her rice cakes, and she always saved some for her poorest neighbors.

One day, she was making rice cakes on the table in her little house, taking handfuls of sticky rice from her pot and pressing the rice into flat circles. All of a sudden, the table shook. "An earthquake!" she mumbled. There are often earthquakes in Japan, so this did not worry her. But before she could catch them, her rice cakes rolled off the table and dropped into a hole in her earthen floor! This bothered her very much and, without thinking, she quickly made the hole larger and dropped down into it herself!

"Come back! Come back!" she cried to her rice cakes. But they rolled merrily along. Suddenly, in the middle of the path ahead of her, there appeared a dreadful Oni! Now, "Oni" is the name the Japanese gave to ogre-like creatures. They were big and hulking, with long claws on their fingers and toes, three eyes, huge teeth, and horns on their heads. Their skin was red or blue or yellow. As you can imagine, the little old lady froze. But then the Oni picked up her rice cakes!

"Hey!" she shouted. "You can't have those! They are mine!"

The Oni looked down at her as it stuffed three rice cakes into its mouth.

"Mmmmm," said the Oni. "These are good. Did you make them?"

"Well, yes, I did," said the little old lady, with a certain degree of pride.

"You come with me," stated the Oni as he grabbed her arm.

It was useless to try to fight against the Oni, so the little old lady didn't even try. The Oni dragged her down corridors of stone and through damp caves until she knew she was lost and would never find her way back home. Finally they came to a great cavern. Its walls were pocketed with caves; and as they entered the cavern, the Oni shouted, "I've found us a cook!"

Oni stuck their heads out of each cave. When they saw the little old woman, they began to shout, "Cook! Cook! Cook!" The Oni showed the little old woman a stove and a pot and a bag of rice. "Put water in pot," he said slowly. "Put one grain rice in pot. Stir with paddle."

"One grain of rice?" said the little old woman. "How can I make rice cakes with one grain of rice?"

"One grain. You stir," replied the Oni, giving her a rice paddle.

So, the little old woman did as she was told. And to her amazement, as she stirred, the one grain of rice became two grains, then four, then sixteen, until the entire pot was filled with rice! When it had cooled, she made rice cakes for the Oni. They loved her rice cakes! They demanded her rice cakes for every meal!

Well, soon the little old woman was tired of this endless work and endless rice cakes. But how could she leave? She had no idea how to get back to the surface of the earth. But she did not despair. Surely there was a solution to her dilemma!

She looked around the cavern. Every day, in order to cook the rice, she had to fill the pot with water. Water! There was a small river flowing through the cavern. She thought and thought. Surely that river came to the surface of the Earth eventually? Yes, it must.

One day, when the Oni were all asleep [for Oni sleep during the day and wake at night], the little old woman quietly took the pot and placed it in the water. Quickly, she jumped in and started paddling with the rice paddle. Off she went! But the river was shallow in places, and the pot dragged along the bottom. The Oni woke up!

Now, Oni cannot cross water, but they can drink a tremendous amount of it. So the Oni knelt down by the river and started to drink. And the river began to shrink! The little old woman was in a panic. But when she looked back, she saw that the fish in the river were flopping around, higher and higher as the water disappeared. She picked up as many fish as she could hold and began to throw them at the Oni. Oni love to eat! When they opened their mouths to catch the fish, the water spilled out and filled the river again.

And away went the Little Old Woman, far down the river. After a while, the river did come to the surface of the earth, right at the bottom of her little hill. She steered the pot over to some rocks and jumped out, pushing the pot back into the water. That way, if the Oni chased after her, they would see the pot far away and run after it, never knowing where she had left it.

Happily, she climbed up her little hill and opened the door of her house. She was very tired after her adventure. But she had gained from all the work she had done for the Oni–for she had come away with the magic paddle. Now, she could make as many rice cakes as she wanted and feed all of her neighbors.

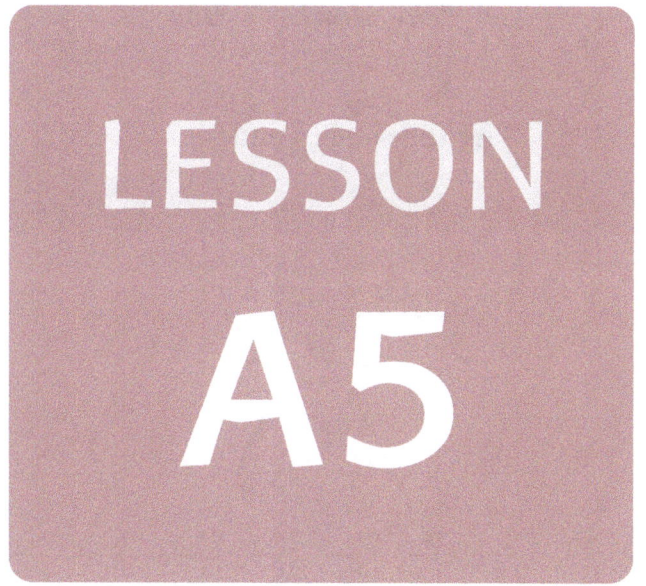

Focus: Continue to explore the possibilities of straight lines; introduction of the circle.

Materials: The same materials as in previous lessons.

Morning Exercises: Repeat in the same order as previously.

Lesson Content:

1. Review *"The Runaway Rice Cakes."* If your child is willing, this is a good story to dramatize. The child can show you how the little old woman runs when she is chasing her rice cakes; how she suddenly freezes when she sees the Oni; how she argues with the Oni but then cooks for them. It's also fun to pretend to be an Oni: to walk like an Oni, chew like one, slurp up a river, show anger at the old woman's escape and chase after her.

2. Now focus on the old woman's stirring the rice, as well as the way she uses the rice paddle as an oar. With the rice paddle, she is making the lines we call "diagonal." Your child will love learning this new word. Draw a diagonal line on the chalkboard.

3. Follow the procedure as above.

4. Open your books to **page 8** and prepare the border as before. Then draw the two diagonal lines / \, not touching.

5. Turn to **page 9**, prepare the border, and say, "This is what will happen when our two diagonal lines meet one another," and draw X.

6. Follow the procedure.

7. Turn to **page 10**, prepare the border as before, and say, "Here is that little rice cake rolling down the path." Draw a large circle on the chalkboard.

8. Follow the procedure.

9. Draw the circle on page 10.

10. Put materials away.

11 Tell *"The Monkey's Heart."*

12 Sing the song to end ML.

Note: Drawing a circle will take some practice, so be sure you work on this form before teaching it! Everyone has a preferred way of drawing a circle. Are you more comfortable drawing a circle in a clockwise or a counter-clockwise direction? Your practice will help you figure this out. You can also draw the circle several times without lifting the chalk from the chalkboard [or crayon from the page]. This will eventually result in a well-formed circle. However, if after much practice, you still cannot draw a circle, you can tell your child, "I've worked on this form for a long time, and it's still hard for me. But I'm going to try it anyway." This courageous attitude will give your child permission not to be perfect. That's an important lesson in itself.

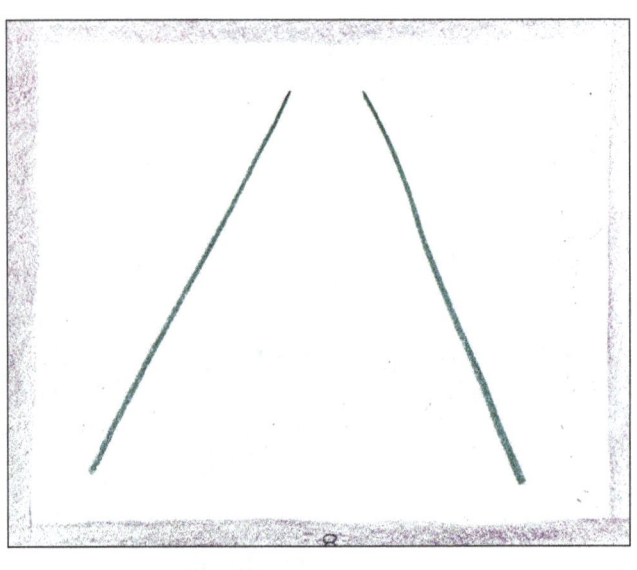

Heart of The Monkey

Lessons A5 and A6

Story in Public Domain

A HUGE TREE LIVED ON THE SEASHORE, half of its branches over the land and the other half over the water. A little monkey lived in the tree. This monkey played in the branches all day and when he was hungry, he ate the sweet fruit that grew in the tree. Now, in the sea there lived a shark. One day the Monkey dropped a piece of fruit into the sea. The shark ate it up.

"Thank you, friend Monkey," said the Shark, "I have only fish to eat in the sea and I like your fruit very much."

The Monkey was happy to be a friend of the Shark and threw fruit into the sea every morning. This went on for some time, until one day, the Shark said to the Monkey, "You are so good to me that I want to do something good for you."

The Monkey looked down at the Shark and listened. The Shark said, "I want to show you my home. You will meet my brothers and sisters. You are so good to me that I think they will like you."

The Monkey thought a minute and said, "Thank you for asking me, but I don't want to go. I am afraid of cold water. And I cannot swim."

"Don't be afraid!" replied the Shark. "Come with me. I shall carry you on my back and I'll swim very slowly."

The Monkey thought, "It is so hot today, and it will be so nice and cool on the water. I think I'll go."

So the Monkey jumped down onto the Shark's back and off they went.

At first the Monkey did not like going on the Shark's back, because the Shark swam very quickly. But soon he relaxed and looked at the new places and at the fish in the water. Everything was so interesting!

"Do you like the sea?" asked the Shark. "Are you glad you came?"

" Yes!" said the Monkey. "How far must we go?"

"My home is not very far," the Shark answered. "But now I must tell you something. Our chief, the biggest shark in the sea, is very ill. Our doctor said to him, 'You must eat a monkey's heart. Then you will be well again.' So I am taking you to him and I am telling it to you, because you are my friend."

The poor Monkey was ready to cry. But he did not cry. The Monkey thought of a plan to save himself. Then he said, "Oh no! Why didn't you tell me that before? I'm so sorry–I left my heart at home, way up in my tree. We Monkeys always hide our hearts in the daytime, and we take them out only at night. I am ready to give my heart to your chief, because I am your friend and I don't want your chief to be angry with you. But how can I do that when my heart is at my home?"

The Shark asked the Monkey, "If I take you back to your tree, will you go and get your heart?"

"Of course, I will. And let us go quickly. Your dear chief must not wait!"

The Shark with the Monkey swam back very quickly. They came again to the big tree. The Monkey climbed up the tree saying, "Wait for me! Wait for me! I'll go get my heart!"

But the monkey did not come back. The Shark was swimming and swimming in the water under the tree. Then he shouted, "Friend Monkey, where are you?"

There was no answer. The Shark thought, "I am afraid he can't find the heart in the branches!"

The Shark waited and waited for the Monkey. Then he shouted again, "Monkey? Monkey? When will you come back to me?"

Again there was no answer. Then the Monkey began to laugh.

"Do you think I am a fool?" asked the little Monkey. "If I give you my heart, I'll die!"

"But you said your heart was in the branches of the tree," said the Shark.

"My heart is where it always is–in my body," shouted the Monkey. "But I will give you some fruit." The monkey tossed a fruit down to the shark. The shark caught the fruit but then swam away and never came back. The little monkey is still laughing.

First Grade Arithmetic—Teacher's Guide

Focus: Exploring the possibilities of the circle.

Materials: The same materials as previously.

Morning Exercises: The same sequence as previously.

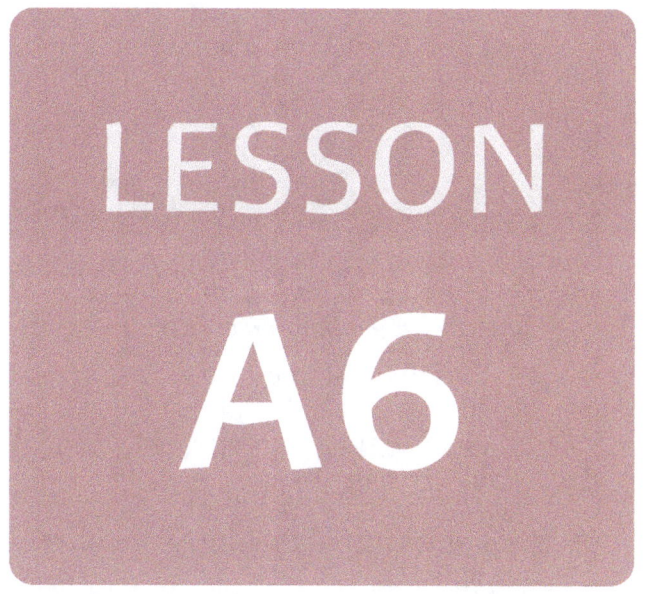

Lesson Content:

1 Review *"The Monkey's Heart."* This story also lends itself to drama. It can be a lot of fun to pretend to be the mischievous monkey or the shark trying to catch the fruit.

2 Say, "Sometimes, when the monkey threw the fruit into the water, it splashed and the water rippled out." Follow the procedure to draw four concentric circles on **page 11**, starting with the smallest circle.

3 Say, "Sometimes, the fruit landed right in the shark's mouth." Follow the procedure and draw four concentric circles with pointed "noses" on **page 12**. Again, start with the smallest circle.

4 Put materials away.

5 Tell *"Turtle's Race with Bear."*

6 Sing the song to end ML.

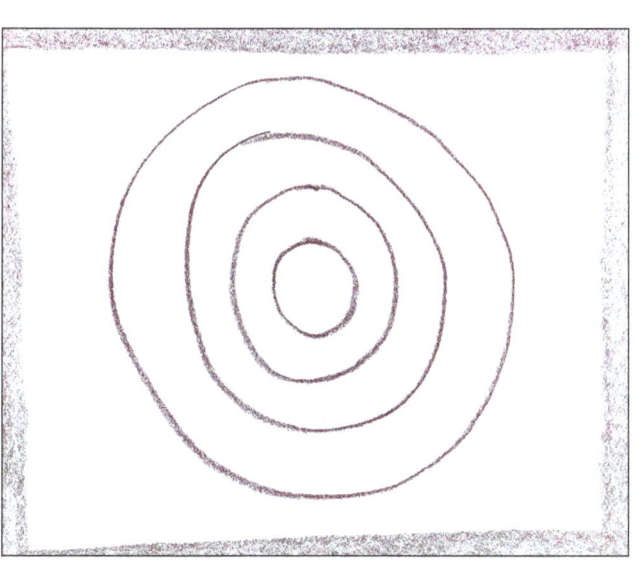

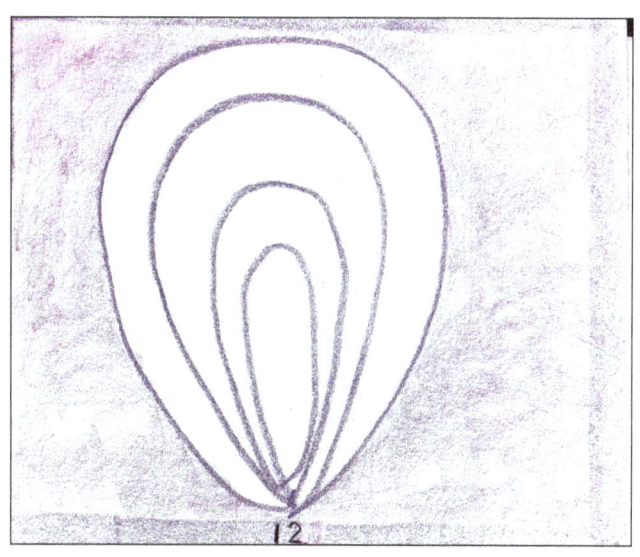

Turtle's Race with Bear

Lessons A6 and A7

Origin: Iroquois. Other cultures have similar legends.
http://www.snowwowl.com/legends/iroquois/iroquois018.html (adapted)
This story can now be found at: www.educationisourbuffalo.com.

IT WAS AN EARLY WINTER, cold enough so that the ice had frozen on all the ponds. And Bear, who had not yet learned in those days that it was wiser to sleep through the Cold Season, grumbled as he walked through the woods.

Perhaps he was remembering a trick another animal had played on him, perhaps he was just not in a good mood. It happened that he came to the edge of a great pond and saw Turtle there with his head sticking out of the ice.

"Hah," shouted Bear, not even giving his old friend a greeting. "What are you looking at, Slow One?"

Turtle looked at Bear. "Why do you call me slow?"

Bear snorted. "You are the slowest of the animals. If I were to race you, I would leave you far behind." Perhaps Bear never heard of Turtle's big race with Beaver and perhaps Bear did not remember that Turtle, like Coyote, is an animal whose greatest speed is in his wits.

"My friend," Turtle said, "let us have a race to see who is the swiftest."

"All right," said Bear. "Where will we race?"

"We will race here at this pond and the race will be tomorrow morning when the sun is the width of one hand above the horizon. You will run along the banks of the pond and I will swim in the water."

"How can that be?" Bear said. "There is ice all over the pond."

"We will do it this way," said Turtle. "I will make holes in the ice along the side of the pond and swim under the water to each hole and stick my head out when I reach it."

"I agree," said Bear. "Tomorrow we will race."

When the next day came, many of the other animals had gathered to watch. They lined the banks of the great pond and watched Bear as he rolled in the snow and jumped up and down making himself ready.

Finally, just as the sun was a hand's width in the sky, Turtle's head popped out of the hole in the ice at the starting line. "Bear," he called, "I am ready."

Bear walked quickly to the starting place and as soon as the signal was given, he rushed forward, snow flying from his feet and his breath making great white clouds above his head. Turtle's head disappeared in the first hole and then in almost no time at all reappeared from the next hole, far ahead of Bear.

"Here I am Bear," Turtle called. "Catch up to me!" And then he was gone again. Bear was astonished and ran even faster. But before he could reach the next hole, he saw Turtle's green head pop out of it.

"Here I am, Bear," Turtle called again. "Catch up to me!" Now bear began to run in earnest. His sides were puffing in and out as he ran and his eyes were becoming bloodshot, but it was no use. Each time, long before he would reach each of the holes, the green head of Turtle would be there ahead of him calling out to him to catch up!

When Bear finally reached the finish line, he was barely able to crawl. Turtle was waiting there for him, surrounded by all the other animals. Bear had lost the race. He dragged himself home in disgrace, so tired that he fell asleep as soon as he reached his home. He was so tired that he slept until the warm breath of Spring came to the woods again.

It was not long after Bear and all the other animals had left the pond that Turtle tapped on the ice with one of his claws. At his sign, a dozen heads like his popped up from the holes all along the edge of the pond. It was Turtle's cousins and brothers, all of whom looked just like him!

"My relatives," Turtle said, "I wish to thank you. Today we have shown Bear that it does not pay to call other people names. We have taught him a good lesson."

Turtle smiled and a dozen other turtles, all just like him, smiled back. "And we have shown the other animals," Turtle said, "that Turtles are not the slowest of the animals."

Origin: Iroquois. Other cultures have similar legends.

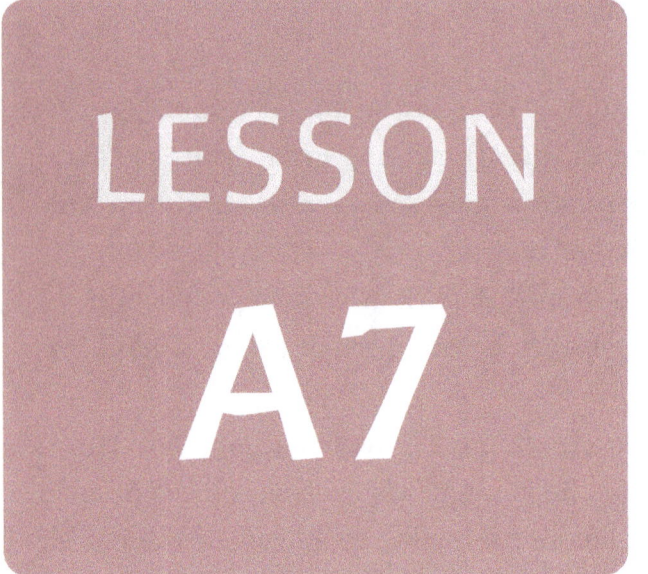

Focus: The first "running form.".

Materials: The same materials as previously, with the addition of a 9" play ball.

Morning Exercises: *[These exercises will be used for the second set of 6 lessons.]*

Through the Night [song]
Through the night, the Angels kept
Watch beside me while I slept.
Now the dark has gone away.
We are glad for this new day.

Morning Has Come [song]
Morning has come, night is away.
We rise with the sun and welcome the day.

Round and Round [song with movement]

Round and round, the Earth is turning,	*[Standing straight and tall, turn slowly in place until you are facing forward at the end of the song.]*
Turning always round to morning,	
And from darkness round to light.	

Tall Trees in the Forest [verse with movement]
[Repeat three times, varying speed of movement once the child knows the verse.]

Tall trees in the forest;	*[Standing straight and tall, lift up on toes and reaching high with your arms.]*
Pine cones on the ground.	*[Slowly bend over and "touch" the pine cones.]*
Tall trees in the forest,	*[Standing straight and tall.]*
They bend, they bend, they bend.	*[Bend over slowly and touch the floor.]*

The Hokey-Pokey [dance song with gestures, as indicated]
You put your hand in, you take your hand out,
You put your hand in and you shake it all about.
Do the hokey-pokey and you turn yourself around:
That's what it's all about!   *[Clap and stomp in rhythm on each syllable.]*

You put your other hand in…..
You put your foot in……
You put your other foot in……

The "dance" can continue for another few verses, but be careful not to spend the entire lesson on the Hokey-Pokey! Other body parts to put in: knee, elbow, back, head. I did not use "left" and "right" because first grade children might be able to identify these at the beginning of the year. However, feel free to use them if your child will not be confused.

Straight as a Spear [verse with movement]
Straight as a spear, I stand.   *[Standing as tall and straight as possible.]*
Strength fills my legs and arms.   *[Spread legs apart, stretch arms straight out to sides.]*
Warmth fills my heart with love.   *[Cross arms over heart.]*
[Swinging your arms around to each side, jump back into upright position, arms at sides.]

Stepping Stones [verse with movement]
          [Step forward on accented syllables. Notice that some steps are slow, some faster.]
Step on **step**ping stones, **one, two, three.**
Step on **step**ping stones, **fol-low me.**
The **river** is **fast,** the **river** is **wide.**
We'll **step** on **step**ping stones **to** the **other side.**

Double, Double *[Try this at varying speeds.]*
Double, double   *[Form hands into fists & thump one on top of the other; switch.]*
This, this.   *[Push the back of each hand forward, in rhythm.]*
Double, double   *[Repeat as above.]*
That, that.   *[Push the front of each hand forward, in rhythm.]*
Double this.   *[One fist bump, then the backs of the hands forward.]*
Double that.   *[The opposite fist bump, then the fronts of the hands forward.]*
Double, double, this, that.   *[Fist bumps, back of hands, front of hands.]*

Count to 10, bouncing a play ball on each number and holding it in your hands on the pause between numbers.

I Am a Tall Tree [verse with movement]

I am a tall tree, reaching up to the sky.	*[Standing straight, stretch arms up.]*
When the wind blows, I lean with a sigh.	
I lean to the front,	*[Legs straight, lean to the front.]*
I lean to the back,	*[Legs straight, lean to the back.]*
I stand up straight without a crack.	*[Jump back into uprightness.]*

The Sun with Loving Light [verse with no movement]

[Stand up straight and tall, crossing arms over the heart, with reverence.]

The Sun with loving light makes bright for me each day.
The Soul with Spirit power gives strength unto my limbs.
In sunlight shining clear, I do revere, O God,
The strength of humankind which you so graciously
Have planted in my soul,
That I with all my might
May love to work and learn.
From you stream light and strength.
To you rise love and thanks.

I Look at My Hands [verse with gestures]

I look at my hands with my fingers fine,	*[Hold hands out in front of you.]*
And I want to be proud that they are mine.	
For deep in my heart lies a golden chest	*[Cross hands over heart.]*
With secret treasures that no one can guess	
Unless my hands do their very best.	*[Hold hands out in front of you.]*

Warm Our Hearts [song to end Morning Lesson]

Warm our hearts, O Sun, and give	*[Cross hands over heart.]*
Light that we may daily live,	
Growing as we want to be:	
True and good and strong and free.	

Lesson Content:

1. Review **Turtle's Race with Bear**. Children tend to love this story, so it lends itself easily to conversation about bragging, playing tricks on people, etc.

2. Ask the child to stand and pretend, with you, to be a turtle under the ice. Take turns poking your noses up into the air.

3. Return to seat. Draw the form you just made in the air in front of you.

4. Open your books to **page 13**. Follow the procedure.

5. Draw the form across the middle of this page.

6. Put materials away. Tell: **The Queen Bee**.

7. Sing the song to end ML.

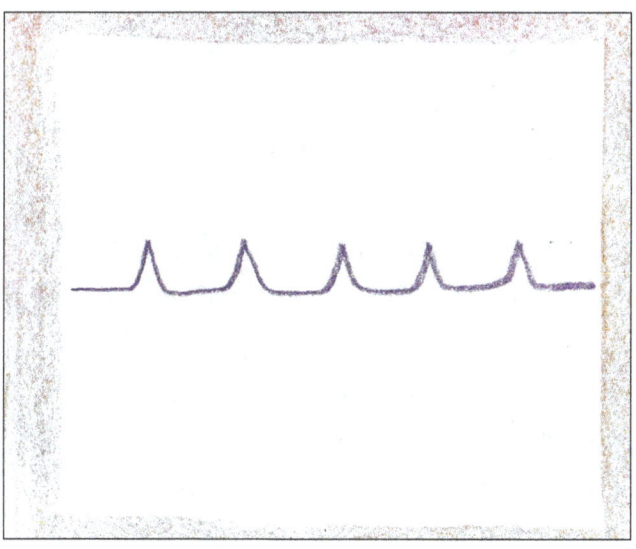

The Queen Bee

Lessons A7, A8, and A9

Story in Public Domain

TWO KING'S SONS once started to seek adventures, and fell into a wild, reckless way of living, and gave up all thoughts of going home again. Their third and youngest brother, who was called Witling, and had remained behind, started off to seek them; and when at last he found them, they jeered at his simplicity in thinking that he could make his way in the world, while they who were so much cleverer were unsuccessful.

But they all three went on together until they came to an ant-hill, which the two eldest brothers wished to stir up, that they might see the little ants hurry about in their fright and carrying off their eggs, but Witling said, "Leave the little creatures alone, I will not suffer them to be disturbed."

And they went on farther until they came to a lake, where a number of ducks were swimming about. The two eldest brothers wanted to catch a couple and cook them, but Witling would not allow it, and said, "Leave the creatures alone, I will not suffer them to be killed." And then they came to a bee's-nest in a tree, and there was so much honey in it that it overflowed and ran down the trunk. The two eldest brothers then wanted to make a fire beneath the tree, that the bees might be stifled by the smoke, and then they could get at the honey. But Witling prevented them, saying, "Leave the little creatures alone, I will not suffer them to be stifled."

At last the three brothers came to a castle where there were in the stables many horses standing, all of stone, and the brothers went through all the rooms until they came to a door at the end secured with three locks, and in the middle of the door a small opening through which they could look into the room. And they saw a little grey-haired man sitting at a table. They called out to him once, twice, and he did not hear, but at the third time he got up, undid the locks, and came out. Without speaking a word he led them to a table loaded with all sorts of good things, and when they had eaten and drunk he showed to each his bed-chamber.

The next morning the little grey man came to the eldest brother, and beckoning him, brought him to a table of stone, on which were written three things directing by what means the castle could be delivered from its enchantment The first thing was, that in the wood under the moss lay the pearls belonging to the princess - a thousand in number - and they were to be sought for and collected, and if he who should undertake the task had not finished it by sunset, if but one pearl were missing, he must be turned to stone. So the eldest brother went out, and searched all day, but at the end of it he had only found one hundred; what was said on the table of stone came to pass and he was turned into stone.

The second brother undertook the adventure the next day, but it fared with him no better than with the first; he found two hundred pearls, and was turned into stone. And so at last it was Witling's turn, and he began to search in the moss; but it was a very tedious business to find the pearls, and he grew so out of heart that he sat down on a stone and began to weep. As he was sitting thus, up came the ant-king with five thousand ants, whose lives had been saved through Witling's pity, and it was not very long before the little insects had collected all the pearls and put them in a heap.

Now the second thing ordered by the table of stone was to get the key of the princess's sleeping-chamber out of the lake. And when Witling came to the lake, the ducks whose lives he had saved came swimming, and dived below, and brought up the key from the bottom.

The third thing that had to be done was the most difficult, and that was to choose out the youngest and loveliest of the three princesses, as they lay sleeping. All bore a perfect resemblance each to the other, and only differed in this, that before they went to sleep each one had eaten a different sweetmeat, the eldest a piece of sugar, the second a little syrup, and the third a spoonful of honey. Now the Queen-bee of those bees that Witling had protected from the fire came at this moment, and trying the lips of all three, settled on those of the one that had eaten honey, and so it was that the king's son knew which to choose.

Then the spell was broken; every one awoke from stony sleep, and took their right form again. And Witling married the youngest and loveliest princess, and became king after her father's death. But his two brothers had to put up with the two other sisters.

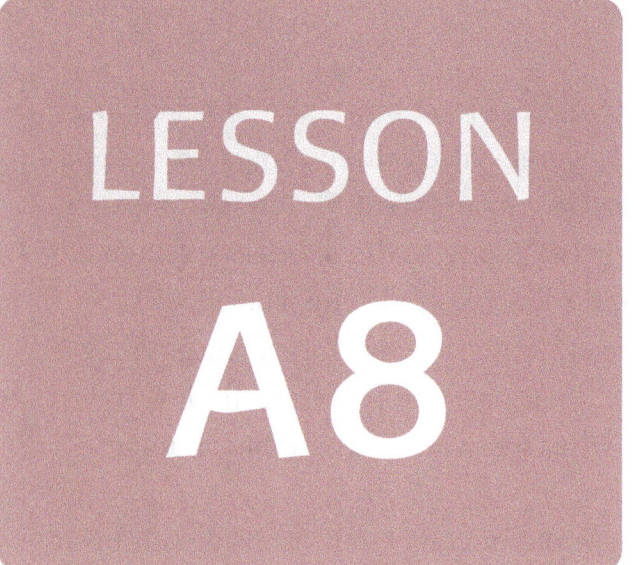

Focus: The illustration for *"The Queen Bee."*

Materials: The same materials as previously.

Morning Exercises: The same exercises as previously.

Lesson Content:

1. Open your books to **page 14.** On this page, you will draw an illustration for *"The Queen Bee."* First, draw the floor of the room with the "papa bear" side of a block crayon. Next, draw a bed for the princess on the left side of the paper. Draw a sleeping princess on the bed, starting with her head.

2. Next, draw the prince standing at the foot of the bed. Start with his head and work down to his feet.

3. With the "mama bear" side of your crayon, color in the wall. Work with a very slight pressure from left to right. Now draw a "queen bee" hovering over the head of the princess.

4. Put materials away.

5. Sing the "song to end ML."

First Grade Arithmetic—Teacher's Guide 33

LESSON A9

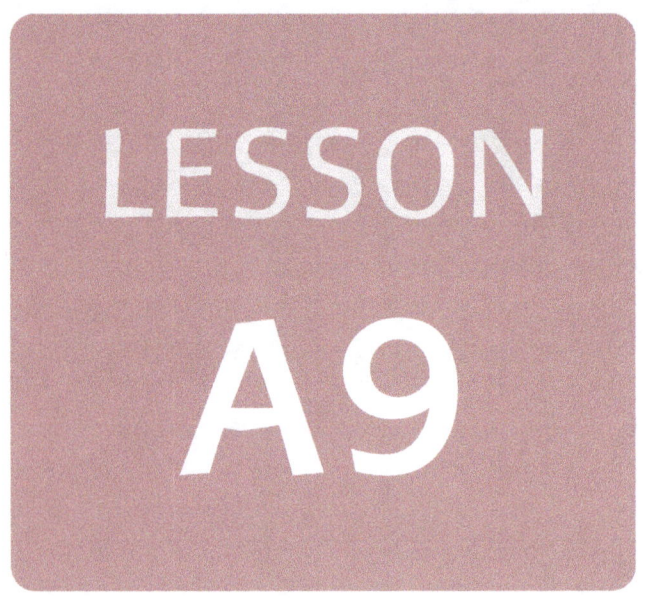

Focus: The spiral.

Materials: The same as previously.

Morning Exercises: The same as previously.

Lesson Content:

1. Briefly review **The Queen Bee.** Open your books to **page 14** and talk about the illustration. This story can lead to a discussion of sweetness, and even to sampling! You can also think about baking later in the day, or even just making toast and trying a couple of sweet toppings.

2. Say, "This little queen bee flew into the room where the princesses were sleeping and hovered over each princess' mouth, tasting the sweetener she had eaten. When the queen bee hovered, she moved like this." Demonstrate how the bee moved in a spiral.

3. Ask the child to stand and move in a large spiral, starting at the outer edge and walking to the center; turning around and then walking back to the outer edge. Moving in a spiral is a lot of fun, especially if you have a large space [inside or outside]. After you walk the form, practice moving it in different ways: running, skipping, jumping

4. Open your books to **page 15**. Follow the procedure and draw this spiral on the page.

5. Put your materials away. Tell: **King Thrushbeard.** [Try to avoid being too dramatic in the telling!]

6. Sing the song to end ML.

King Thrushbeard

Lessons A9 and A10

Story in Public Domain

A KING HAD A DAUGHTER who was beautiful beyond all measure, but so proud and haughty that no suitor was good enough for her. She sent away one after the other, and ridiculed them as well.

Once the King made a great feast and invited, from far and near, all the young men likely to marry. They were all marshalled in a row according to their rank and standing; first came the kings, then the grand-dukes, then the princes, the earls, the barons, and the gentry. Then the King's daughter was led through the ranks, but to every one she had some objection to make; one was too fat, "The wine-cask," she said. Another was too tall, "Long and thin has little in." The third was too short, "Short and thick is never quick." The fourth was too pale, "As pale as death." The fifth too red, "A fighting-cock." The sixth was not straight enough, "A green log dried behind the stove." So she had something to say against every one, but she made herself especially merry over a good king who stood quite high up in the row, and whose chin had grown a little crooked. "Well," she cried and laughed, "he has a chin like a thrush's beak!" and from that time he got the name of King Thrushbeard.

But the old King, when he saw that his daugher did nothing but mock the people, and despised all the suitors who were gathered there, was very angry, and swore that she should have for her husband the very first beggar that came to his doors.

A few days afterwards a fiddler came and sang beneath the windows, trying to earn a small alms. When the King heard him he said, "Let him come up." So the fiddler came in, in his dirty, ragged clothes, and sang before the King and his daughter, and when he had ended he asked for a trifling gift. The King said, "Your song has pleased me so well that I will give you my daughter there, to wife."

The King's daughter shuddered, but the King said, "I have taken an oath to give you to the very first beggar-man, and I will keep it." All she could say was in vain; the priest was brought, and she had to let herself be wedded to the fiddler on the spot. When that was done the King said, "Now it is not proper for you, a beggar-woman, to stay any longer in my palace, you may just go away with your husband."

The beggar-man led her out by the hand, and she was obliged to walk away on foot with him. When they came to a large forest she asked, "To whom does that beautiful forest belong?" – "It belongs to King Thrushbeard; if you had taken him, it would have been yours." – "Ah, unhappy girl that I am, if I had but taken King Thrushbeard!"

Afterwards they came to a meadow, and she asked again, "To whom does this beautiful green meadow belong?" – "It belongs to King Thrushbeard; if you had taken him, it would have been yours." – "Ah, unhappy girl that I am, if I had but taken King Thrushbeard!" Then they came to a large town, and she asked again, "To whom does this fine large town belong?" – "It belongs to King Thrushbeard; if you had taken him, it would have been yours." – "Ah, unhappy girl that I am, if I had but taken King Thrushbeard!"

"It does not please me," said the fiddler, "to hear you always wishing for another husband; am I not good enough for you?" At last they came to a very little hut, and she said, "Oh goodness! what a small house; to whom does this miserable, mean hovel belong?" The fiddler answered, "That is my house and yours, where we shall live together."

She had to stoop in order to go in at the low door. "Where are the servants?" said the King's daughter. "What servants?" answered the beggar-man; "you must yourself do what you wish to have done. Just make a fire at once, and set on water to cook my supper, I am quite tired." But the King's daughter knew nothing about lighting fires or cooking, and the beggar-man had to lend a hand himself to get anything fairly done. When they had finished their scanty meal they went to bed; but he forced her to get up quite early in the morning in order to look after the house.

For a few days they lived in this way as well as might be, and came to the end of all their provisions. Then the man said, "Wife, we cannot go on any longer eating and drinking here and earning nothing. You weave baskets." He went out, cut some willows, and brought them home. Then she began to weave, but the tough willows wounded her delicate hands.

"I see that this will not do," said the man; "you had better spin, perhaps you can do that better." She sat down and tried to spin, but the hard thread soon cut her soft fingers so that the blood ran down. "See," said the man, "you are fit for no sort of work; I have made a bad bargain with you. Now I will try to make a business with pots and earthenware; you must sit in the market-place and sell the ware." – "Alas," thought she, "if any of the people from my father's kingdom come to the market and see me sitting there, selling, how they will mock me?" But it was of no use, she had to yield unless she chose to die of hunger.

For the first time she succeeded well, for the people were glad to buy the woman's wares because she was good-looking, and they paid her what she asked; many even gave her the money and left the pots with her as well. So they lived on what she had earned as long as it lasted, then the husband bought a lot of new crockery. With this she sat down at the corner of the market-place, and set it out round about her ready for sale. But suddenly there came a drunken hussar galloping along, and he rode right amongst the pots so that they were all broken into a thousand bits. She began to weep, and she did not know what to do for fear. "Alas! what will happen to me?" cried she; "what will my husband say to this?"

She ran home and told him of the misfortune. "Who would seat herself at a corner of the market-place with crockery?" said the man; "leave off crying, I see very well that you cannot do any ordinary work, so I have been to our King's palace and have asked whether they cannot find a place for a kitchen-maid, and they have promised me to take you; in that way you will get your food for nothing."

The King's daughter was now a kitchen-maid, and had to be at the cook's beck and call, and do the dirtiest work. In both her pockets she fastened a little jar, in which she took home her share of the leavings, and upon this they lived.

It happened that the wedding of the King's eldest son was to be celebrated, so the poor woman went up and placed herself by the door of the hall to look on. When all the candles were lit, and people, each more beautiful than the other, entered, and all was full of pomp and splendour, she thought of her lot with a sad heart, and cursed the pride and haughtiness which had humbled her and brought her to so great poverty.

The smell of the delicious dishes which were being taken in and out reached her, and now and then the servants threw her a few morsels of them: these she put in her jars to take home.

All at once the King's son entered, clothed in velvet and silk, with gold chains about his neck. And when he saw the beautiful woman standing by the door he seized her by the hand, and would have danced with her; but she refused and shrank with fear, for she saw that it was King Thrushbeard, her suitor whom she had driven away with scorn. Her struggles were of no avail, he drew her into the hall; but the string by which her pockets were hung broke, the pots fell down, the soup ran out, and the scraps were scattered all about. And when the people saw it, there arose general laughter and derision, and she was so ashamed that she would rather have been a thousand fathoms below the ground.

She sprang to the door and would have run away, but on the stairs a man caught her and brought her back; and when she looked at him it was King Thrushbeard again. He said to her kindly, "Do not be afraid, I and the fiddler who has been living with you in that wretched hovel are one. For love of you I disguised myself so; and I also was the hussar who rode through your crockery. This was all done to humble your proud spirit, and to punish you for the insolence with which you mocked me."

Then she wept bitterly and said, "I have done great wrong, and am not worthy to be your wife." But he said, "Be comforted, the evil days are past; now we will celebrate our wedding." Then the maids-in-waiting came and put on her the most splendid clothing, and her father and his whole court came and wished her happiness in her marriage with King Thrushbeard, and the joy now began in earnest. I wish you and I had been there too.

Focus: The story: *King Thrushbeard.*

Materials: Same as previously.

Morning Exercises: Same as previously.

Lesson Content:

1 Review the story *King Thrushbeard*. This is often one of the children's favorite stories, as they seem to love the descriptions of the various noble suitors. Conversations about the story will naturally include questions about the cruelty of the princess, but also about the behavior of her father and of her "beggar" husband. What's fair?

2 Open your books to **page 16.** Draw the illustration of the suitors.

3 Put your materials away.

4 Sing the song to end ML.

LESSON A11

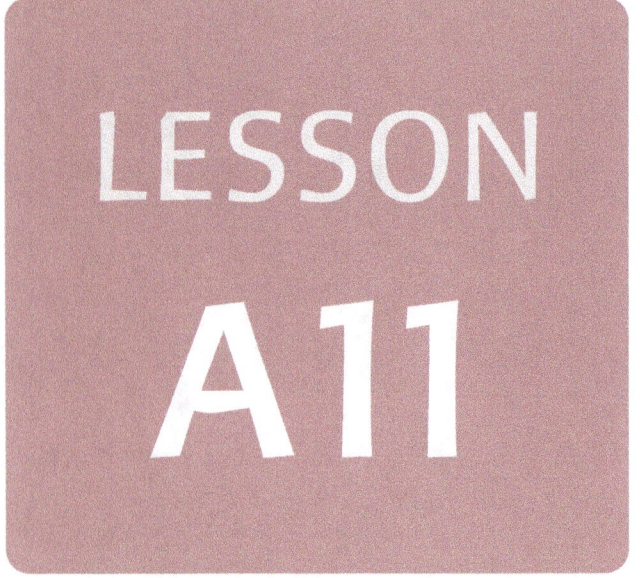

Focus: Drawing vertical lines of varying heights. The story: *The Princess and the Pea*.

Materials: The same as previously.

Morning Exercises: The same as previously.

Lesson Content:

1. Briefly review the story of **King Thrushbeard,** this time focusing on the different heights of the suitors. You might talk about the different heights of people or pets you know.

2. Turn to **page 17.** Follow the procedure and draw the form.

3. Put the materials away.

4. Tell: *The Princess and the Pea.*

5. Sing the song to end ML.

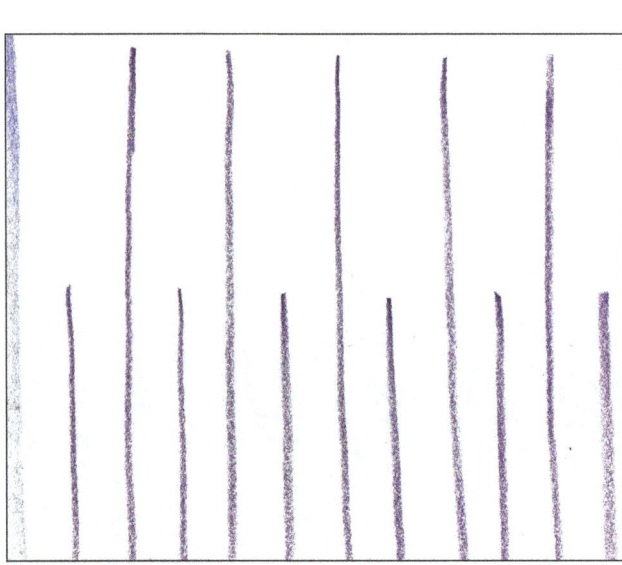

The Princess and the Pea

Lessons A11 and A12

By Hans Christian Andersen 1835

THERE WAS, ONCE UPON A TIME, a prince who wanted to marry a princess, but he was determined only to marry a true princess. So he traveled through the whole world to find one, but there was always something against each. There were plenty of princesses, but he could not find out if they were true princesses. In every case there was some little defect, which showed that the princess was not a true princess. So he came home again in very low spirits. One night there was a dreadful storm; there was thunder and lightning, and the rain streamed down in torrents. It was fearful! There was a knocking heard at the palace gate, and the old king went to open it.

There stood a princess outside the gate; but oh! What a sad plight she was in from the rain and the storm! The water was running down from her hair and her dress into the points of her shoes and out at the heels again. Yet she said she was a true princess.

"Well, we shall soon find that out!" thought the old queen. But she said nothing, and went into the sleeping room, took off all the bedclothes, and laid a pea on the bottom of the bed. Then she put twenty mattresses on top of the pea, and twenty heavy quilts on the top of the mattresses. This was the bed in which the princess was to sleep.

The next morning she was asked how she had slept.

"Oh, very badly!" said the princess. "I scarcely closed my eyes all night! I am sure I don't know what was in the bed. I laid on something so hard that my whole body is black and blue. It was dreadful!"

Now they perceived that she was a true princess, because she had felt the pea through the twenty mattresses and the twenty heavy quilts. No one but a true princess could be so sensitive.

So the prince married her, for now he knew that at last he had gotten hold of a true princess. The pea was put into the Royal Museum, where it is still to be seen, if no one has stolen it. Now this is a true story.

LESSON A12

Focus: Horizontal lines of the same length, separated by the same distance.

Materials: The same as previously.

Morning Exercises: The same as previously.

Lesson Content:

1. Review *The Princess and the Pea.*

2. Open your books to **page 18**. Follow the procedure and draw the form.

3. Close your books and put them away.

4. Sing the song to end ML.

In the Waldorf schedule for First Grade, blocks of Arithmetic alternate with Language Arts blocks. When you have completed Lesson A12 with your child, we recommend that you start a four-week study of consonants. In the first Language Arts block, you should be able to introduce eight – ten of the twenty-one consonants.

Along with your Language Arts lessons, we offer Skills Classes. We recommend that you teach Language Arts three times per week in the same rhythm that you've followed for the Arithmetic classes. Twice-weekly Skills Classes can be added in at your convenience. For example, you could teach Skills Classes on the days without Language Arts lessons, at the same time you usually teach Language Arts. This schedule would create a five-day rhythm for your child. There are other possibilities for scheduling the Skills Classes—think about what will work best for your family. ■

MAIN LESSON BLOCK B

The Quality of Numbers

RESEARCH HAS SHOWN that most children count automatically from a very young age, even if they only count numbers one, two, and three in correct order. From the beginning of the year, we count as part of the *morning exercises*. Counting is always accompanied with movement and speech: step, march, jump, clap, and combine any of these movements to create the pattern you need. Children will be encouraged to count on their fingers and toes–how much more personal can we make arithmetic than to use our own bodies?

Counting is so automatic, in fact, that as soon as we know how to count a few numbers, we are often encouraged to jump right in to adding them together. We seldom stop to consider what any specific number means, however, and Waldorf Education considers meaning to be the foundation of learning any skill and gaining knowledge of any kind.

What, in fact, is ONE? What is "one-ness?" Ask a class of children to name something in the world of which there is only one, and chances are several will respond that each person is a "one." And they're right! This can lead into an exciting and revealing conversation centered around the question: What is special about you? Why are you unique? Moreover, we each have TWO hands; and with those two hands, we can join with two other people to make THREE. The foundations of arithmetic are thus found in the human being; numbers have meaning and significance for the child far beyond a counting exercise. Steiner instructed teachers to *"Seize every opportunity to use images and tangible objects. This helps children find a way into the real world and to form everything in keeping with reality."* [Human Values in Education]

In this course, I've used similar examples when discussing the numbers in class, and examples from the natural world when illustrating in our ML books: What in Nature is only ONE? What does all life depend on? Someone will realize that I am talking about the Sun, and that is what we'll draw in our books. We'll draw a different illustration for each number from 1 - 12. Our number system is a "base ten" number system; but TWELVE is what we might call an iconic number. There are twelve signs of the Zodiac, twelve months of the year. Twelve is also a number that is easily factored for all four operations of arithmetic.

Practice in subitizing is also part of *morning exercises* throughout the year. "Subitizing" refers to the ability to automatically recognize a quantity and associate it with its numerical value. The ability to subitize is foundational for further work with numbers. For example, when you see three dots on a piece of paper, do you have to stop and count them, or do you automatically know there are three dots? Most children and adults can recognize up to 5 items without counting them.

As Steiner, as well as many mathematicians, has pointed out, understanding mathematics resides within the body. So when introducing a mathematical concept to children we include as many senses as we can. Touch, movement, balance, vision, hearing–our senses provide us with an almost unlimited resource for teaching arithmetic. When introducing the written numbers, for example, think of various ways to incorporate movement and touch. Instruct the child to "draw" the number in the air and walk the form of the number on the floor. Walk a number yourself while the child watches and then ask the child to guess which number you're making. Use the sense of touch by "drawing" a number on the child's back and asking them to guess which number you're making. Clap a certain number of times and ask the child to tell you how many times you clapped. These are only a few examples of activities you can bring to your child in solidifying their understanding of number. [Please remember that some numbers are easily written backward, so watch carefully to ensure that your child is writing them correctly.]

Each lesson asks that you use two pages for that particular number: one page for an illustration representing the number, and one page for practicing the number. We advise that you finish the illustration within the lesson. However, if your child does not finish practicing writing the number within the lesson, plan to come back to that page and practice more later. Some children like to hurry through any assignment, but it is very important that they draw each number carefully and correctly. The horizontal bars of color, created by the widest side of the block crayon, are intentionally large. Some children tend to write small, but the entire bar of color should be used in forming the numbers. This is developmentally appropriate for first grade students.

Materials:
- Spiral ML book #2
- Block and stick crayons
- Drawing mat
- 2" beanbag
- Chalkboard, chalk, and eraser
- 9" play ball

Focus: The number ONE.

Materials:
- Main Lesson Book #2
- Block and stick crayons
- Large pencils
- 9" play ball

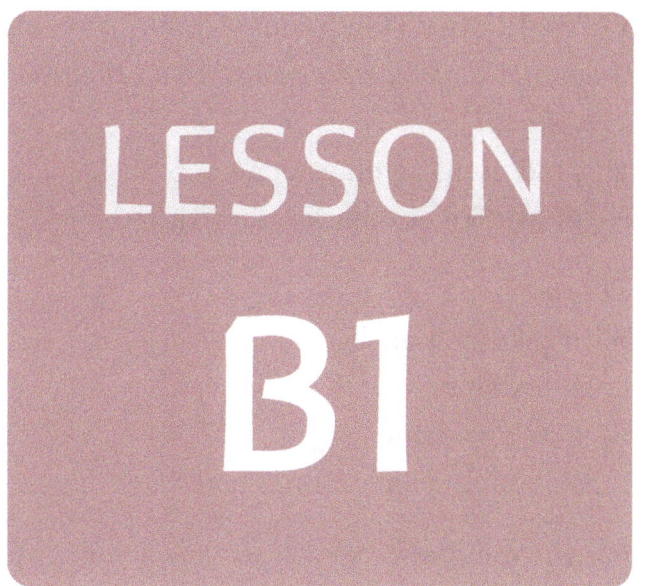

Morning Exercises:

Through the Night [song]
Through the night, the Angels kept
Watch beside me while I slept.
Now the dark has gone away.
We are glad for this new day.

Morning Has Come [song]
Morning has come, night is away.
We rise with the sun and welcome the day.

Round and Round [song with movement]

Round and round, the Earth is turning,   *[Standing straight and tall, turn slowly in*
Turning always round to morning,   *place until you are facing forward at the*
And from darkness round to light.   *end of the song.]*

Tall Trees in the Forest [verse with movement]
[Repeat three times, varying speed of movement once the child knows the verse.]

Tall trees in the forest;   *[Standing straight and tall, lift up on toes*
   and reach high with your arms.]

Pine cones on the ground.   *[Slowly bend over and "touch" the pine cones.]*
Tall trees in the forest,   *[Standing straight and tall.]*
They bend, they bend, they bend.   *[Bend over slowly and touch the floor.]*

This Old Man [Traditional Children's Song]
[I usually play this game sitting on the floor, but you can play it standing or sitting in a chair.]

This old man, he played one.	*[Hold up one finger.]*
He played knick-knack on my thumb.	*[Point to thumb.]*

Chorus:

With a knick-knack	*[Clap on "knick-knack."]*
Paddy-whack	*[Pat knees on "paddy-whack."]*
Give a dog a bone	*[Pretend to throw a bone.]*
This old man came rolling home.	*[Roll hands over lap while swaying side to side.]*

For the remaining verses, use the same gestures. Sing the chorus after each verse.

This old man, he played two.
He played knick-knack on my shoe.

This old man, he played three.
He played knick-knack on my knee.

This old man, he played four.	
He played knick-knack on my door.	*[Hold one hand up, palm flat, as the "door."]*

This old man, he played five.	
He played knick-knack on a beehive.	*[Hold up one fist as the "beehive."]*

This old man, he played six.	
He played knick-knack on some sticks.	*[Splay fingers and cross one hand over the other.]*

This old man, he played seven.	
He played knick-knack up to Heaven.	*[Point upwards.]*

This old man, he played eight.	
He played knick-knack on my gate.	*[Hold up one hand, palm flat, & swing back and forth.]*

This old man, he played nine.	
He played knick-knack up a vine.	*[Pretend to climb a vine.]*

This old man, he played ten.	
He played knick-knack all over again.	*[With hands and face, show dismay!]*

You Are One [by Vivian Jones-Schmidt

You are **one** and **I** am **one**,   *[Bounce ball on accented syllables.]*
And **one** is the **gol**den **Sun**.

Golden **Sun** and **Moon** make **two**,
And **so do I** and **you**.

Sun and **Moon** and **Earth** will **be**
Always to**geth**er and **al**ways **three**.

Fall and **Winter**, **add** two **more**:
Spring and **Sum**mer make **sea**sons **four**.

I am a **star** that **shines** so **bright**,
With **five** strong **points** that **make** my **light**.

Count to ten and back to one, bouncing the playball on each bounce and holding it in your hands in the pause between numbers.

I Am a Tall Tree [verse with movement]
I am a tall tree, reaching up to the sky.   *[Standing straight, stretch arms up.]*
When the wind blows, I lean with a sigh.
I lean to the front,   *[Legs straight, lean to the front.]*
I lean to the back,   *[Legs straight, lean to the back.]*
I stand up straight without a crack.   *[Jump back into uprightness.]*

Straight as a Spear [verse with movement]
Straight as a spear, I stand.   *[Standing as tall and straight as possible.]*
Strength fills my legs and arms.   *[Spread legs apart, stretch arms straight out to sides.]*

Warmth fills my heart with love.   *[Cross arms over heart.]*
[Swinging your arms around to each side, jump back into upright position, arms at sides.]

The Sun with Loving Light [verse with no movement]
[Stand up straight and tall, crossing arms over the heart, with reverence.]
The Sun with loving light makes bright for me each day.
The Soul with Spirit power gives strength unto my limbs.
In sunlight shining clear, I do revere, O God,
The strength of humankind which you so graciously
Have planted in my soul,
That I with all my might
May love to work and learn.
From you stream light and strength.
To you rise love and thanks.

I Look at My Hands [verse with gestures]
I look at my hands with my fingers fine,	*[Hold hands out in front of you.]*
And I want to be proud that they are mine.	
For deep in my heart lies a golden chest	*[Cross hands over heart.]*
With secret treasures that no one can guess	
Unless my hands do their very best.	*[Hold hands out in front of you.]*

Warm Our Hearts [song to end Morning Lesson]
Warm our hearts, O Sun, and give	*[Cross hands over heart.]*
Light that we may daily live,	
Growing as we want to be:	
True and good and strong and free.	

Lesson Content:

1 Lead the class through the conversation described in the introductory section. If the response "the Sun" is not forthcoming, say," When I think of one thing in Nature that we can't live without, I think of the Sun. There is only one Sun, high in the sky even when it's covered with clouds."

2 Turn to **page 2** in your book. [You will use page 1 as the title page for this book.] Draw a gloriously large, golden Sun in the center of the page. Surround the Sun with a blue sky.

3 Turn to **page 3**. With the large, "papa bear" side of a block crayon, draw a horizontal line left to right across the top of the page. Choose another, complimentary, color and draw a line directly underneath this line. Continue to alternate colors of lines until the page is full. [For example, you might use yellow and blue. It often helps to use a light color and a darker color for contrast.]

4 Make "ones" in the air as in Form Drawing. Then practice writing simplified, vertical, "ones" across the page on the top line. Draw "ones" on every other line until you fill the page. [You will be drawing on the same colored lines.]

5 Close your books and put the materials away. Say, "Today we talked about the number ONE, and we drew a Sun. Tomorrow we will talk about the number TWO. Tonight, I want you to think about what we might draw that is TWO."

6 Sing the song to end ML.

LESSON B2

Focus: The numbers TWO and THREE.

Materials: As in Lesson B1.

Morning Exercises: As in Lesson B1.

Lesson Content:

1. Open your books to **pages 2 & 3.** Review your conversation about the number ONE.

2. Now ask, "We drew the Sun to represent the number ONE. What do you think we could add to the Sun so that, together, they make TWO?" If your child does not think of "the Moon," guide the conversation so that the Moon is the obvious addition to the Sun.

3. Open your books to **page 4.** Draw a golden Sun in the upper left hand corner and a yellow Moon in the lower right hand corner. Imagine a diagonal line from upper right to lower left on the page. With the medium side of the blue crayon, draw blue around the Sun; with the medium side of the purple crayon, draw purple around the Moon. You can also use the golden stick crayon to draw rays coming from the Sun, and the yellow stick crayon to draw curved lines [representing the "glow" of the Moon] around the Moon.

4. On **page 5,** choose two colors of block crayons to draw alternating lines as in the previous lesson.

5. Practice drawing a simple TWO in the air as in Form Drawing. Now draw TWO's on alternating bars of color.

6 Now say, "We have ONE as the Sun; and TWO as the Sun and the Moon. What could we add to make THREE?" The response we are looking for is "the Earth." On **page 6,** draw a large Sun in the upper left; a smaller Moon on the right, midway down the page; and a still smaller Earth in the lower center of the page.

7 On **page 7,** draw alternating bars of color as before. Trace the number THREE in the air. Then write it on alternate bars of color on the page.

8 Say, "In our next lesson, we'll talk about the next number, which is four. Try to bring something to class that has four parts."

9 Close your books and put the materials away.

10 Sing the song to end ML.

LESSON B3

Focus: The number FOUR.

Materials: As before.

Morning Exercises: As before.

Lesson Content:

1. Review the conversations for the numbers ONE, TWO, and THREE.

2. Now talk about the number FOUR. Ask your child to hold up four fingers. What can you think of that has or is FOUR? If you have a dog or cat, of course the animal has four legs! But if you have a fruit tree nearby, you can also talk about what the tree looks like throughout the year. Talk about the seasons.

3. Open your books to **page 8.** Show the child how to divide the page into four sections, measuring with block crayons. First, lay 8 block crayons across the top of the page, long side down. Count to the middle, make a mark with a pencil, and remove the crayons. Now draw a line straight down the middle of the page. This is a very good form drawing exercise! Now, repeat the exercise, laying the crayons down the side of the page and drawing a horizontal line across.

4. Discuss which season to start with. Draw a tree in that season, in the upper left square. Proceed through all three squares, left to right, top to bottom, drawing the tree in each season.

5. On **page 9,** draw horizontal bars of alternating colors. Practice drawing the FOUR in the air; then draw it on alternating bars on the page.

6. Close your books and put the materials away.

7. Sing the song to end ML.

Focus: The number FIVE.

Materials: As before.

Morning Exercises: As before.

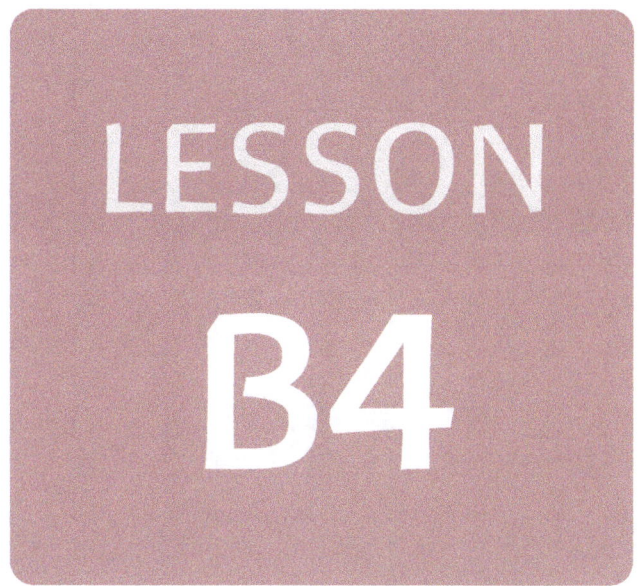

LESSON B4

Lesson Content:

1. Say, "Today we'll talk about the number FIVE. Can you think of anything that is FIVE?" The child will probably think of five fingers or five toes. Great! Now say, "Can you make FIVE with your whole body?" If the child doesn't think of how to do this, show them. Standing straight and tall, spread your arms and legs out and ask, "Do you see the five in me?" "Can you make five with your body? What do we look like when we make FIVE?" You look like stars!

2. Open your books to **page 10.** With a golden stick crayon, draw an image of a human being with arms and legs outstretched. Please do not draw this as a stick figure, but rather as a full figure. Now, with a yellow stick crayon, draw a five-pointed star around the figure.

3. On **page 11,** draw alternating bars of color. Trace a FIVE in the air, walk it on the floor, etc. Now write in on alternating lines on the page.

4. Close the books and put away the materials.

5. Sing the song to end ML.

LESSON B5

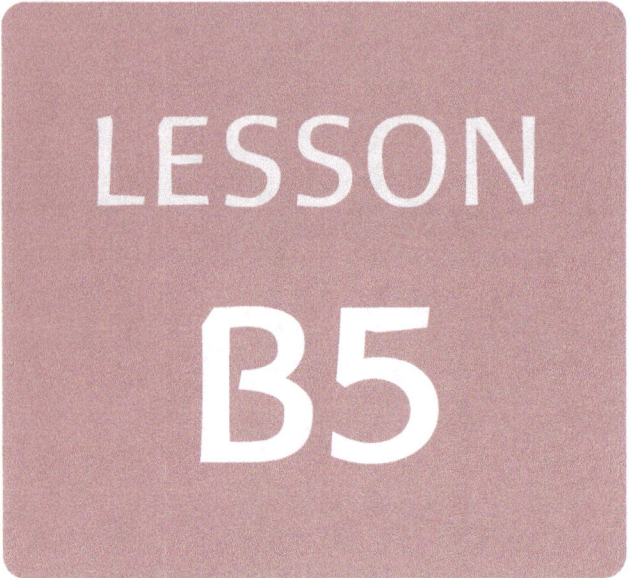

Focus: The number SIX. If at all possible, bring a honeycomb to class.

Materials: As before.

Morning Exercises: As before.

Lesson Content:

1. If you've been able to locate a honeycomb, show it to your child and marvel at the golden color and the perfection of the form. Point to one cell and ask your child to count the sides of the cell. There should be six sides!

2. Talk about the little bees, how hard they work keeping their house clean and flying out to find flowers, bringing pollen back to the hive where they make honey and wax. Talk about your beeswax crayons and how grateful we are to the bees for their hard work.

3. Open your books to **page 12.** Show how to draw a hexagon in the air and practice tracing it in the air with your child. Now draw it on your page with a golden stick crayon. This can be a difficult form to draw, so your child might need help [and you will need to practice ahead of time]. I've found it easiest to draw two parallel lines for the top and bottom first; then place dots diagonally out to each side and connect them to the ends of the two parallel lines.

4. Fill the cell with golden crayon. Draw a little bee outside the cell; then shade the page with blue.

5. Turn to **page 13** and draw horizontal lines of alternating colors. Practice "drawing" the SIX in the air and then write it on every other line of the page.

6. Close your books, put materials away.

7. Sing the song to end ML.

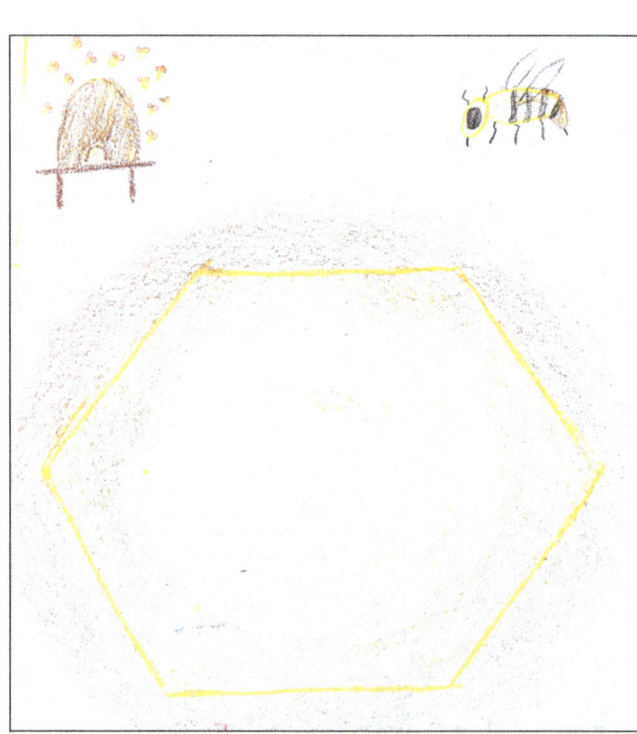

LESSON B6

Focus: The number SEVEN.

Materials: As before.

Morning Exercises: As before.

Lesson Content:

1. Say, "Today we're going to talk about the number SEVEN. This is another special number. What can you think of that is SEVEN?" If your child doesn't immediately say, "The rainbow!" give them hints: We sometimes see it in the sky. Especially after a rainstorm.

2. Now we're going to draw a rainbow and put in all of the colors. Which color comes first? Make sure you've thought about this beforehand: starting at the top, red, orange, yellow, green, blue, indigo, violet. [ROY G BIV]

3. Open your books to **page 14.** Draw a large rainbow on your page. Your child can also draw a house under it, or a person, flowers, trees, etc.

4. Trace a SEVEN in the air with your child. On **page 15,** draw horizontal lines of alternating colors as before. Practice writing SEVENS on every other line.

5. Close your books. Put away all materials.

6. Sing the song to end ML.

LESSON B7

Focus: The number EIGHT.

Materials: As before, with the addition of a 2" bean bag.

Morning Exercises:

Through the Night [song]
Through the night, the Angels kept
Watch beside me while I slept.
Now the dark has gone away.
We are glad for this new day.

Morning Has Come [song]
Morning has come, night is away.
We rise with the sun and welcome the day.

Round and Round [song with movement]

Round and round, the Earth is turning,	[*Standing straight and tall, turn slowly in place until you are facing forward at the end of the song.*]
Turning always round to morning,	
And from darkness round to light.	

Tall Trees in the Forest [verse with movement]
[*Repeat three times, varying speed of movement once the child knows the verse.*]

Tall trees in the forest;	[*Standing straight and tall, lift up on toes and reach high with your arms.*]
Pine cones on the ground.	[*Slowly bend over and "touch" the pine cones.*]
Tall trees in the forest,	[*Standing straight and tall.*]
They bend, they bend, they bend.	[*Bend over slowly and touch the floor.*]

One, Two, Buckle My Shoe [with a play ball]
[Bounce the ball on each accented syllable, and catch on the unaccented syllable.]
One, **two,** buckle my **shoe.**
Three, **four,** shut the **door.**
Five, **six,** pick up **sticks.**
Seven, **eight,** lay them **straight.**
Nine, **ten,** a big fat **hen.**
Eleven, **twelve,** dig and **delve.**
Thirteen, **fourteen,** dishes need **sorting.**
Fifteen, **sixteen,** take them to the **kitchen.**
Seventeen, **eighteen,** we're all **waiting.**
Nineteen, **twenty,** there's work a-plenty.

Count to twenty, walking forward one step on each number. Now count back from twenty to one, taking one step backward on each number. Steps can be very small and very loud!

Bean bag Exercises
[There are seven bean bag exercises in the sequence, but we will start with one. Each exercise is accompanied by a poem.]

Cross-patch
[Hold the bean bag in your right hand, chin-height above the left hand, which is under the right hand and at waist level. Drop the bean bag into the left hand on the accented syllable. Now switch places with the hands. Drop the bean bag from the left hand [on top] into the right hand [underneath] on the accented syllable. Switch hands again and continue to drop the bean bag into the lower hand on the accented syllable.]
Cross patch, **draw** the latch.
Sit by the fire and **spin.**
Take a cup and **fill** it up.
In**vi**te the neighbors **in.**

I Am a Tall Tree [verse with movement]

I am a tall tree, reaching up to the sky.	*[Standing straight, stretch arms up.]*
When the wind blows, I lean with a sigh.	
I lean to the front,	*[Legs straight, lean to the front.]*
I lean to the back,	*[Legs straight, lean to the back.]*
I stand up straight without a crack.	*[Jump back into uprightness.]*

Straight as a Spear [verse with movement]

Straight as a spear, I stand.	*[Standing as tall and straight as possible.]*
Strength fills my legs and arms.	*[Spread legs apart, stretch arms straight out to sides.]*
Warmth fills my heart with love.	*[Cross arms over heart.]*

[Swinging your arms around to each side, jump back into upright position, arms at sides.]

The Sun with Loving Light [verse with no movement]

[Stand up straight and tall, crossing arms over the heart, with reverence.]

The Sun with loving light makes bright for me each day.
The Soul with Spirit power gives strength unto my limbs.
In sunlight shining clear, I do revere, O God,
The strength of humankind which you so graciously
Have planted in my soul,
That I with all my might
May love to work and learn.
From you stream light and strength.
To you rise love and thanks.

I Look at My Hands [verse with gestures]

I look at my hands with my fingers fine,	*[Hold hands out in front of you.]*
And I want to be proud that they are mine.	
For deep in my heart lies a golden chest	*[Cross hands over heart.]*
With secret treasures that no one can guess	
Unless my hands do their very best.	*[Hold hands out in front of you.]*

Warm Our Hearts [song to end Morning Lesson]

Warm our hearts, O Sun, and give	*[Cross hands over heart.]*
Light that we may daily live,	
Growing as we want to be:	
True and good and strong and free.	

Lesson Content:

1 Review the numbers you've worked with thus far. Ask your child to hold up the correct number of fingers as you say the numbers. Say them in order forwards and backwards, and then in a random order.

2 Now say, "Today we are talking about the number **eight.** Can you show me eight fingers? Do you know of anything in our world that is **eight** of something that we can draw?"

3 Continue to talk until you arrive at the realization that a spider has eight legs. Turn to **page 16** in your book. Draw the web, first drawing a vertical line down the center of the page; then a horizontal line across the middle of the page; then a diagonal line from upper left to lower right; then a diagonal line from upper right to lower left. These lines should all cross in the center of the page. Now draw the spider, with four legs on each side, over the center.

4 On **page 17**, draw horizontal bars of color as before. Trace the **eight** in the air. This is a good number to "walk" on the floor! When you have practiced the **eight** in several modalities, write it on every other line on **page 17.**

5 Close your books and put your materials away.

6 Sing the song to end ML.

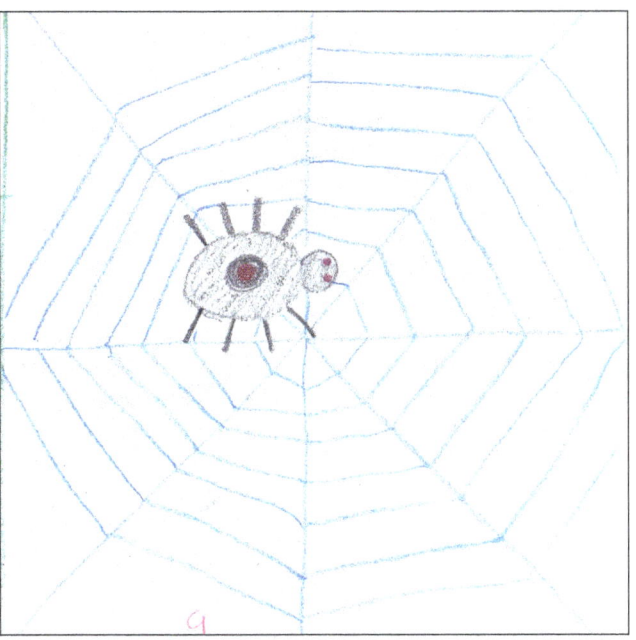

LESSON B8

Focus: The number NINE.

Materials: As before.

Morning exercises: As before.

Lesson Content

1. The number 9 is challenging. I have not been able to find anything in Nature that can be referred to as having nine parts. Therefore, I always talk about candelabras and, as this block often falls close to Hannukah, I mention the menorah with its nine candles. If you are not familiar with this observance, you can still talk about candelabras. It is also fun to make your own out of clay or wood.

2. You can draw the candelabra in any form you wish. Here is how I've always drawn it:

a. Turn to **page 18.** Draw a yellow "helping line" from top to bottom in the center of the page.
b. With the large side of the crayon, draw a bar of color from left to right along the bottom of the page. This is your "table top."
c. Draw the base of the candelabra sitting on top of the "table" and extending about 3" upward. I often use a golden stick crayon for this.
d. Now draw a line across the top and extending about an inch in each direction. Draw "candles" in the center and at each end.
e. Extend the horizontal line another inch or so in each direction, and draw candles at the ends. Continue doing this until you have a center candle and four candles on each side.
f. With the middle side of a block crayon, draw swirls of color for the background "wall."

3. Now draw horizontal lines of alternating colors on **page 19** and practice the 9.

4. Close your books and put away your materials. Sing the song to end ML.

First Grade Arithmetic—Teacher's Guide

Focus: The number TEN.

Materials: As before.

Morning Exercises: As before.

Lesson Content:

1. Say, "We've talked about a lot of numbers!" Ask your child to show you the fingers for all of the numbers, 1 - 9, in order, backwards, and randomly.

2. Now say, "What number comes next?" They will probably know that the correct answer is "ten." "Show me **ten** fingers. Look at that! We have ten fingers! What else do we have ten of? Yes! We have ten toes!"

3. "Today we're going to draw both of our hands, showing our ten fingers. Open your book to **page 20**. Place both hands on your page and move them around until you can get both hands on the page without blocking any fingers. My hands are larger than yours, so I'll have to really work at this!"

4. When you have figured out how to place your hands, trace around one hand at a time, using a colored pencil. Your child can decorate their hands in any way they want– rings? Nail polish? Tattoo?

5. Now draw horizontal lines of alternating colors on **page 21** and practice writing the **10**. It is a good idea to draw a dot or a star between each **10** to separate the numbers.

6. As an additional treat, take off your shoes and trace your feet on a large sheet of paper, with all ten toes evident.

7. Close your books, put away your materials, and sing the song to end ML.

LESSON B10

Focus: The number ELEVEN.

Materials: As before.

Morning Exercises: As before.

Lesson Content:

1. Say, "We've talked about a lot of numbers, haven't we? Do you remember the picture we drew for the number ONE? What did we draw for the number FOUR? What did we draw for the number NINE? Now we come to the next number. What is that number?"

2. When you've agreed that it is ELEVEN, say," Children who are in First Grade usually need ELEVEN hours of sleep every night. We're going to use the beautiful colors of our crayons to represent those eleven hours. Open your book to **page 22**."

3. Blending any two colors of block crayons, draw the floor of a child's room with the large side of the crayons. Next, draw a child resting in a bed on the floor in the center of the page.

4. You will be drawing the colors in a fan shape around the bed; the colors will mirror one another. With a blue block crayon on the medium side, draw a bar of color extending from the bed to either side of the page. On top of the blue, draw purple, then other colors in this order: magenta, red, orange. In the middle, creating a cone shape over the child, use gold. Starting on the left and counting over to the right, you will have drawn 11 bars of color.

5. On **page 23,** draw horizontal lines of alternating colors and practice writing the **11**. Again, stars or dots between the elevens are very helpful.

6. Close your books, put away your materials, and sing the song to end ML.

Focus: The first six illustrations for the number TWELVE.

Materials: As before.

Morning Exercises: As before.

Lesson Content:

1 To illustrate the number **twelve,** we will be considering the twelve months of the year. In this lesson, we'll complete illustrations for at least the first six months. You will want to adjust the illustrations to reflect the months and seasons in your area, as these illustrations arose out of conversations with children who live in the mid-Atlantic area of North America.

2 Review all of the numbers you have discussed thus far, again using fingers and toes. You might, for example, count those ten toes and add one finger. Now add one more finger and count to twelve twice. Some children have difficulty remembering the names "eleven" and "twelve," so repetition is important.

3 Ask your child what they know of that has twelve components, and bring the conversation around to twelve months of the year. Starting with January, recite the twelve months with your child.

4 Now say, "We're going to draw a little illustration for each of these months, but before we can begin to draw, we need to make twelve little boxes on our page. Open your book to **page 24**."

5 You will be making a grid on this page, with three columns of squares across and four rows of squares down. You'll need your block crayons for measuring the squares.

6 Lay three block crayons, long side down, end to end, starting at the top left side of the paper. At the end of the third crayon, make a little mark on the page. Now lay the crayons, long side down, end to end, starting at the top right side of the paper. At the end of the third crayon, make a little mark. Using a stick crayon, draw the straightest lines you can from your little marks to the bottom of the page. You have divided the page into three columns.

7 Now lay eleven of your block crayons on the left side of the page, stacked on top of one another on their long sides. Be sure the crayons are stacked tightly and that the top of the top crayon is flush with the top of the page. Count three crayons down and make a little mark. Make marks after the next three crayons; and then after the next three. You should have three marks on the left side of the page, dividing your page into four rows. Repeat this sequence on the right side of the page.

8 Now draw the straightest lines you can, left to right, linking the marks on the left side with the mirror marks on the right side. You now have twelve almost-squares on your page. Moreover, the twelve divided into four rows of three create a row for each season.

9 Start with Spring on the top row. Talk about each month as you decide which illustrations to draw. Sometimes your child will follow your lead on how to illustrate a particular month; sometimes your child will have their own ideas. In my class, children drew birthday cakes for the months of their birthdays, so each child's page was unique in at least that way.

10 Spring: March, daffodils; April, a robin with a nest; May, a rabbit.

11 Summer: June, a sailboat; July, a picnic; August, a sand bucket and shovel.

12 Six illustrations is probably all you will be able to accomplish in this lesson. Plan to finish the page in the next lesson.

13 Close your books, put away your materials, and sing the song to end ML.

LESSON B12

Focus: The final six illustrations for the number TWELVE.

Materials: As before.

Morning Exercises: As before.

Lesson Content:

1. Open your books and review the previous lesson.

2. Continue illustrating the number TWELVE.

3. Autumn: September, apple tree; October, tree with multicolored leaves; November, pumpkins.

4. Winter: December, Christmas tree; January, snowman; February, Valentines.

5. Now review all twelve months.

6. On **page 25,** draw horizontal bars of alternating colors and practice writing the number TWELVE. Again, a star or a dot between numbers will help to keep them separated.

7. Close your books, put away your materials, and sing the song to end ML.

You have now completed two arithmetic Morning Lesson blocks. In the Waldorf curriculum, your next Morning Lessons will be in Language Arts, focusing on more consonants. In a four week block, you should be able to introduce 8 - 10 of the 21 consonants in the English alphabet. You will have more Language Arts blocks during the academic year, in which you will be able to introduce any remaining consonants as well as all of the vowels.

In this curriculum, we now provide a block of supportive arithmetic skills classes to accompany your next Language Arts block. As before, these skills class lessons are meant to be taught twice each week at a time of your convenience. ■

MAIN LESSON BLOCK C

Introducing the Four Operations

THIS BLOCK BEGINS WITH the Roman Numerals, thus providing the children with a review of the numbers 1 - 12 as presented in the previous block. A simple story introduces the historical need for symbols in counting and the development of the Roman Numeral symbols from the use of the fingers for counting. Once these "finger symbols" have been introduced, play with your child by asking them to show you Roman Numbers between one and twelve with their fingers.

The four basic operations of arithmetic are addition, subtraction, multiplication, and division, and they're usually taught in that order, over the first three grades. The Common Core Curriculum, for example, recommends teaching addition and subtraction in Grades K-2, and multiplication and division in Grades 3-5. In the Waldorf curriculum, we teach all four operations in the same set of lessons from First Grade on. Initially, we work with numbers such as 12 and 24, which are easy to decompose for multiplication [4 x 3, 2 x 12] and division [12 divided by 2, 24 divided by 8] as well as for addition and subtraction. By teaching all four operations "at once," the children are encouraged to think of numbers as being flexible, mobile, and versatile, rather than fixed. This approach, combined with our practice of starting with the whole number and then proceeding to the parts, supports the development of a certain freedom in thinking. Moreover, research has revealed that "high-achieving math students don't just know more....they engage in flexible thinking when they work with numbers, decomposing and recomposing numbers." [Boaler, *What's Math Got To Do With It?* page 141.] These are critical skills for success in arithmetic as well as in higher mathematics. For example, consider 12:

12 = 2 + 10	12 = 3 x 4	12 divided by 2 = 6
12 = 5 + 7	12 - 8 = 4	12 divided by 3 = 4
12 - 6 = 6	12 = 6 x 2	12 divided by 12 = 1

Twelve is a number of almost limitless possibilities!

Another unique aspect of Waldorf Education is the fact that we work only with whole numbers in Grades 1 - 3. Experience and observation has confirmed that this approach allows the child time to form a solid foundation in these four operations before introducing fractions.

As was stated in the Introduction, it is important as well as interesting to consider the moral qualities that can be seen as underlying the four operations. In teaching addition and/or multiplication, for example, are we teaching acquisitiveness or greed? Does the way we teach subtraction or division reinforce attachment to material goods? Can we think of ways to teach these operations that highlight generosity, compassion, interest in other people? This theme will appear again as each operation is discussed in the text.

In Steiner's first lectures to the teachers, he connected the four operations with the four temperaments [See 'The Temperaments' in the Appendix.] Addition was discussed as being a *phlegmatic* exercise; subtraction was seen as *melancholic*; multiplication was seen as *sanguine*; and division as *choleric*.

Traditionally, Waldorf teachers have created little gnome characters for each operation and used these gnomes in stories about working with numbers. In this course, I've chosen not to use gnomes, which have their own important part to play in the natural world. Rather, I've introduced children of various temperaments to represent the operations. The temperaments are associated with certain colors, and those are the colors I've used for the operations. In the first block, we work with operations on numbers 1 - 12, always starting concretely with fingers, toes, and counting stones; proceeding to the imaginative quality of illustrations; and finally working with abstract numbers.

Thus far, we have focused on numbers as qualities and considered iconic matches for each number quality. Now, we begin to work with the interactions of numbers, learning how to manipulate them as quantities. Waldorf education is an education that nurtures and relies upon the imagination, in arithmetic as in other subjects. So, we start from the imagination in teaching the manipulation of numbers: we start with stories. Some teachers create stories with gnomes as the characters; some use human characters. This curriculum begins with four children in the block that introduces the operations [Lesson Block C] and brings in gnomes in the final arithmetic block of the year [Lesson Block D]. Lesson Block C focuses on numbers 1 - 12, thus allowing the children time to develop an understanding of the four operations before expecting them to work with larger numbers, which are harder to imagine.

The language of arithmetic is very important. When young students learn only that the "equals sign" points to "the answer" in an equation, they learn that the "equals sign" is an operational sign rather than a sign that describes a relationship. So for example, many students are unable to correctly complete the equation: $5 + 7 = \_\_\_\_ + 10$. Rather than giving the correct response of "2," they fill in the blank with 12. Two ways to reinforce the relational meaning of the equals sign are:

1. Replace "equals" with "is the same as" often as you read equations.
2. Use examples, like the one above, of the "equal sign" being used to demonstrate relationships.

While perhaps not so relevant in First Grade arithmetic, be aware of this issue and work to change your language at least from time to time. As you go through the grades, focus on this more and more. When your child begins to study Algebra and algebraic thinking, they will thank you.

For this block, you will be using 12 counting stones [or buttons, rocks, beans, etc.]. It will be most helpful if you and your child keep your counters in a container or a little bag–anything unspillable.

LESSON C1

Focus: Roman Numerals.

Materials: As before.

Morning Exercises:

Through the Night [song]
Through the night, the Angels kept
Watch beside me while I slept.
Now the dark has gone away.
We are glad for this new day.

Morning Has Come [song]
Morning has come, night is away.
We rise with the sun and welcome the day.

Round and Round [song with movement]

Round and round, the Earth is turning,	[Standing straight and tall, turn slowly
Turning always round to morning,	in place until you are facing forward the
And from darkness round to light.	at end of the song.]

Tall Trees in the Forest [verse with movement]
[Repeat three times, varying speed of movement once the child knows the verse.]

Tall trees in the forest;	[Standing straight and tall, lift up on toes and reach high with your arms.]
Pine cones on the ground.	[Slowly bend over and "touch" the pine cones.]
Tall trees in the forest,	[Standing straight and tall.]
They bend, they bend, they bend.	[Bend over slowly and touch the floor.]

Clap Hands [traditional children's song]

Clap, clap, clap your hands,   *Do the motions as you sing the words.]*
Clap your hands together.
Clap, clap, clap your hands,
Clap your hands together.

Stomp your feet.....

Nod your head.....

Jump up high.....

Turn around.....

[Add any motions you or the child think of!]

Bean Bags

Cross-patch, draw the latch,   *[motions as before]*
Sit by the fire and spin.
Take a cup and fill it up,
Invite the neighbors in.

Whisky, frisky, hippety-hop!   *[Start with the bean bag held in front*
Up he goes, to the tree top!   *and toss it around your body in time*
Whirly, twirly, round and round,   *with the verse.]*
Down he scampers to the ground.
Furly, curly, what a tail!
Tall as a feather, broad as a sail.
Where's his supper? In the shell.
Snappy-cracky, out it fell.

One Shoe, Two Shoes

One shoe, two shoes   *[1. Tap left foot with right hand in front,*
    then right foot with left hand, also in front.]
Three shoes, four.   *[2. Tap left foot with right hand, behind;*
    then right foot with left hand, also behind.]
Check our pony's silver shoes,   *[Pattern #1.]*
Let's tap them all once more.   *[Pattern #2.]*

One shoe, two shoes,
Three shoes, four.
Let's go riding, pony dear,
Through the stable door.

[Repeat as above.]

Clippety-Clop
[Pretend you are horses, trotting through the countryside. Follow the directions.
Clip-clop, clippety-clop,
To the countryside, don't stop!
Clip-clop, clippety-clop,
Faster, faster—slower, slower–
Whew! It's time to stop!

Counting with Play Ball
Hold the ball in your hands on the odd numbers;
Bounce the ball on the even numbers and speak louder.
Count to 12; stop. Count back from 12 to 1.

Do this several times, varying the speed and also the volume of your voice.
For example, try whispering on the odd numbers and shouting on the even numbers.

When your child is comfortable with the numbers 1 - 12, keep going to 24, forward and back.

I Am a Tall Tree [verse with movement]
I am a tall tree, reaching up to the sky.	*[Standing straight, stretch arms up.]*
When the wind blows, I lean with a sigh.	
I lean to the front,	*[Legs straight, lean to the front.]*
I lean to the back,	*[Legs straight, lean to the back.]*
I stand up straight without a crack.	*[Jump back into uprightness.]*

Straight as a Spear [verse with movement]
Straight as a spear, I stand.	*[Standing as tall and straight as possible.]*
Strength fills my legs and arms.	*[Spread legs apart, stretch arms straight out to sides.]*
Warmth fills my heart with love.	*[Cross arms over heart.]*

[Swinging your arms around to each side, jump back into upright position, arms at sides.]

The Sun with Loving Light [verse with no movement]
[Stand up straight and tall, crossing arms over the heart, with reverence.]
The Sun with loving light makes bright for me each day.
The Soul with Spirit power gives strength unto my limbs.
In sunlight shining clear, I do revere, O God,
The strength of humankind which you so graciously
Have planted in my soul,
That I with all my might
May love to work and learn.
From you stream light and strength.
To you rise love and thanks.

I Look at My Hands [verse with gestures]

I look at my hands with my fingers fine,	*[Hold hands out in front of you.]*
And I want to be proud that they are mine.	
For deep in my heart lies a golden chest	*[Cross hands over heart.]*
With secret treasures that no one can guess	
Unless my hands do their very best.	*[Hold hands out in front of you.]*

Warm Our Hearts [song to end Morning Lesson]

Warm our hearts, O Sun, and give	*[Cross hands over heart.]*
Light that we may daily live,	
Growing as we want to be:	
True and good and strong and free.	

Lesson Content:

1 **Story:**
"Counting the Sheep"

A long time ago, there lived a family with two children, a girl and a boy. Everyone in the family worked so that they would all have food and clothes and a home to live in. The parents took care of the cow and the donkey and the pigs, and their fields and garden. The children took care of their twelve sheep. Every day they led the sheep from their enclosure near the house out to the grassy fields for grazing. Every evening they led the sheep back to the paddock, where they would be safe.

Every morning and every evening, the children had to count the sheep to make sure they had all twelve. This was a real challenge, because the children were very young. They did not have any paper or pencils, and the sheep kept moving around. How could they make sure they had counted the correct number of sheep? How could they make sure no sheep had been lost or left behind?

Finally the children realized that they had the tools they needed to count correctly—they had their fingers! So, here is the way they counted. [Hold up your fingers as you count for your child.] One finger, two fingers, three fingers, four fingers, five fingers. But wait, five fingers is an entire hand. What's another way to show five? They made a "V" with their thumb and forefinger for the five.

Now on to the other hand. For these numbers, we'll use the "V" to show five. Six fingers, seven fingers, eight fingers, nine fingers, ten fingers. But wait–is there a way we can show ten without using all of our fingers? Yes, we can make two "V's" with our two hands. But the children decided to cross their hands like this [make an "X" with your hands] to show ten.

What number comes after ten? Eleven. To show eleven, they crossed their hands for 10 and then put up one finger. To show twelve, they crossed their hands for 10 and then put up 2 fingers. Now let's count to twelve using our fingers like the children in our story did. [Practice counting from 1 - 12 with fingers several times.]

These are the numbers written long ago by a people we call "the Romans." We call the numbers "Roman numerals." A "numeral" is another word for number. Review the story from yesterday. Practice making your Roman numerals with your fingers again. Call out numbers randomly and ask your child to show you how to "say" them with Roman numerals.

2 Now open your book to **page 26.** The first thing you will need to do is to divide your page into blocks for the numbers. Your page will have four columns vertically and six rows horizontally. Place eight block crayons, on their sides, across the top of your page. Make a mark after each two blocks. You will have three marks, thus dividing the page into four columns. Draw the straightest lines you can from these marks to the bottom of the page. Approach this task as you would a form drawing–practice three times in the air for each line before drawing it.

3 Now place eleven block crayons, on their sides, down the left-hand side of your page. After each two crayons, make a mark. You will have five marks, thus dividing the page into six rows. [The last row will probably be a little shorter.] Draw the straightest lines you can across your page, from left to right. Practice each line three times in the air before drawing it.

4 Now you have a page with 24 little rectangles. You'll be working with two for each number. Across the top of the page in the first row, draw the Arabic [traditional] numerals: 1, 2, 3, and 4. Underneath each, you will draw its Roman equivalent: I, II, III. When you come to 4, explain to your child that the Romans decided that drawing 4 lines was just too many. Ask how else you might draw 4. There might be a little conversation here; but do explain that as 4 is "one before five," the Romans drew a four like this: IV.

5 On the third row, draw the Arabic numerals: 5, 6, 7, 8. Underneath each, draw its Roman equivalent: V, VI, VII, VIII.

6 On the fifth row, draw the Arabic numerals: 9, 10, 11, 12. Underneath each, draw its Roman equivalent. For 9, once again explain how the Romans used a shorter way to write that quantity. Ask your child if they can figure out what the Romans did. They will probably remember how to write "4" in Roman numerals, so the Roman "9" will be easy to figure out. Here are the Roman numerals for row six: IX, X, XI, XII.

7 Now review the Arabic numerals and their Roman equivalents, using your fingers. Say the numbers in mixed order and ask your child to show you each number with their fingers.

8 Close your books, put away your materials, and sing the song to end ML.

LESSON C2

Focus: Review of numbers 1 - 24.

Materials:
- ML book #2
- Block and stick crayons
- Large pencils
- Play ball
- Bean bag

Morning Exercises: As before.

Lesson Content:

1 First, practice counting to 24. Stand and walk around the room, one step for each number. Vary your counting by stomping, jumping, hopping. Whisper or shout the numbers–just don't lose your place!

2 Now return to your seats. Open your books to **page 27.** With the wide side of a block crayon, draw a bar of color across the top of the page, left to right. Immediately under this bar, draw a bar of a different color. Continue to draw bars of color until you have filled the page.

3 On the top line, write the numbers 1 - 6, leaving plenty of room between the numbers. A star between each two numbers separates them nicely. Be sure to write your numbers so that they touch the top and the bottom of the color bar. Writing numbers slowly, carefully, and correctly is important. Make them beautiful in their simplicity.

4 Skip a color bar and write the numbers 7 - 12 on the next bar.

5 Skip a color bar and write the numbers 13 - 18 on the next bar.

6 Skip a color bar and write the numbers 19 - 24 on the last bar.

7 Close your books and put away your materials. Sing the song to end ML.

Focus: Introduction to "Number Land."

Materials: As before.

Morning Exercises: As before.

Lesson Content:

1 Say, "Today I'm going to tell you about a special place called Number Land. In Number Land, everything is the shape of a number: the houses, the trees and flowers, the rocks, the mountains, *everything*. Now of course, the people in Number Land were not shaped like numbers! But all of the play structures on the playground where the children played did look like numbers. What structures do you think you would see on a playground?" [Your child will think of swingsets, sand boxes, slides, etc.]

2 "Let's open our books to **page 28** and draw a picture of that playground." Again, start by drawing the earth across the bottom of the page, left to right, with the widest side of the brown and green block crayons.

3 Now draw some of the play structures, etc., that your child has thought of. *[See the illustration for this lesson as an example.]*

4 Tomorrow we'll learn about some of the children who live in Number Land.

5 Close your books, put away your materials, and sing the song to end ML.

LESSON C4

Focus: Introducing the four operations.

Materials: As before; also 12 counting stones [or buttons, etc.]

Morning Exercises: As before.

Lesson Content:

1. "First, open your bag and take out your stones. It is very important that you always have 12 stones, so count them as you take them out of your bag." Demonstrate how to count the stones, one by one, and place them in a row.

2. "Now, let's count them backwards." Demonstrate how to start at 12 and remove one stone at a time. Place the stones you remove in a separate pile until there are no stones left in your row.

3. Instruct your child to count the stones forward, into a row, and backwards, into a pile, until you are satisfied that the child can do this without your help.

4. "Now count out all twelve of your stones. Separate five on the left with a space in your row. How many are on the other side? So we say, 'twelve is the same as five plus seven.'"

5. "What if you put three before your space? How many are on the other side? Now we say,'Twelve is the same as three plus nine.'"

6. Continue to work with the stones for another three or four combinations, always starting with 12 when you name the equation. Ask your child to name the equations also.

7. "You have twelve stones. Suppose the cat sits on three. How many can you see?" You can use a bowl to be "the cat." It's best to be as visual as possible! "We say, 'twelve minus/take away three is the same as nine.'"

8. Continue to work with examples of the cat's mischief for another three or four equations, always citing the equation and asking your child to name it as well.

9. "Now let's say that you have three friends, and you want to share your stones with them, giving each friend the exact same amount. How many stones will you give each friend?" Demonstrate making a place on the desk/table for each friend. One thing that can work well is to use a blank index card to represent each friend. "We can say, '12 divided by 3 is the same as 4.'" Continue to work in this way, asking your child how many stones they will give each friend if the number of friends is 2, 4, 12.

10. "Now what if you want to give each of your friends 2 stones. How many friends can you share 2 stones with?" Demonstrate making little piles of 2 stones on the desk/table. Count the piles. "We can say, '12 is the same as 2 times 6.'"

11. Continue to work in this way, asking your child how many friends they can

12. "We have done a lot of work with our little stones today. We have added them together and subtracted them, and we have learned how to divide and multiply them. Tomorrow we'll start meeting friends who will help us remember how to do all this work with our stones."

13. "Now let's count our stones again. How many should we have? Yes, if none of our stones have gone on a journey, we should have 12."

14. Put your materials away. Sing the song to end ML.

LESSON C5

Focus: Introducing Division.

Materials: As before.]

Morning Exercises: As before.

Lesson Content:

1. "In our last lesson, we moved our stones in so many different ways. What do you remember about something we did with our stones?" Wait for your child to respond. Encourage them to remember as much about the previous lesson as possible.

2. "Today we're going to meet a friend who will help us work with our stones. This friend is named Divide. He loves to share whatever he has with his neighbors, and he always gives each neighbor exactly the same amount of whatever he has to share: apples, muffins, beautiful stones." [Use examples that are meaningful to your child.] In fact, when he finds a bag of bird seed, he puts the same number of seeds in each hand and holds his hands out for the birds! And that's how we're going to draw him today.

3. Open your books to **page 29**. First, lay four block crayons across the top of your page, widest side down. Make a pencil mark after the fourth crayon. This mark will divide the page into approximately two equal sections. Now take a stick crayon and trace the straightest line you can, three times in the air, from the top to the bottom of the page. Then draw it on the page.

4. With your brown and green block crayons, draw the earth from the left side of the page to the vertical line you just drew. [You will draw an "earth" using two vertical bars of color.]

5 "Choose a brown to draw his face and hands. Draw his face like I draw mine." Until your child is comfortable with drawing people, it will be a good idea for you to place your finger on your child's page in the place for the head.

6 With your red stick crayon, draw the trunk and limbs of Little Divide, arms and hands [drawn in your chosen skin color] outstretched. Draw the feet, then hair and a cap. Draw a "divide" sign on his shirt.

7 Draw six yellow bird seeds in each hand, and a blue bird above each hand.

8 Draw a bright sun above Little Divide and swirls of blue for the sky.

9 "This is Little Divide. In our next class, he will show us how he works."

10 Close your books, put away your materials, and sing the song to end ML.

Focus: Writing the division algorithm.

Materials: As before.]

Morning Exercises: As before.

Lesson Content:

1. Open your books to **page 29.** "What is the name of this helper?" Your child will almost certainly remember the name, but if they don't, say, "This friend is called Little Divide, and he helps us when we have things to share."

2. On the right hand side of the vertical line you drew in the previous lesson, place twelve block crayons, on their sides, from the top of the page to the bottom. Draw a little line under each three crayons. Remove the crayons and extend your marks to the right side of the paper. You have divided that side of the paper into four work spaces.

3. "Take your stones out of their pouch, counting them one by one. How many do you have altogether?" If your child does not have 12, remove the extras or replace the ones that are missing.

4. "Little Divide has a bowl of 12 apples. He wants to give the same number of apples to each of his three friends. How many apples will he give to each friend?"

5. Encourage your child to figure this out independently, using the same approach you used in the previous lesson. Bring out the index cards if that will help.

6. "Now I'm going to teach you how to write this story in numbers." Across the top of the first work space, draw three ovals–one for each "friend."

7. "Now watch. I have 12 apples and 3 friends. First I give each friend one apple." Draw one large dot in each oval. "Now I give each friend another apple." Draw another dot in each oval.

8. Repeat this two more times until you have four large dots in each oval.

9 "Now we'll write this in numbers like grown-ups do. How many apples do we have altogether?" "Yes, we have 12, so that is the first number we write." Write the twelve. "We need a special sign to show that we're giving each friend the same number of apples. This is the sign, and whenever we see this sign, we say 'divided by.'" Draw the division sign after the 12.

10 "Now we need to show how many friends we're giving the apples to, so we write a 3." "Let's read our arithmetic sentence so far: 'Twelve divided by three.' The next sign we need to write will tell us that we know how many apples to give each friend." Write the equals sign. "When we see this sign, we say 'equals' or 'is the same as.' How many apples did we give each friend?"

11 "We gave each friend four apples. Now let's read our arithmetic sentence: 'Twelve divided by three equals four.'"

12 Repeat this process for the following number stories, writing the arithmetic sentences in the work spaces as you complete each story:

a. 12 apples divided by 6 friends equals 2 apples for each friend
b. 12 apples divided by 4 friends equals 3 apples for each friend
c. 12 apples divided by 12 friends equals 1 apple for each friend

13 [You could also use the story: 12 apples divided by 2 friends. You will need four stories for **page 29;** the extra story will be used in the review in the next lesson.]

14 Practice reading aloud the equations/number stories.

15 Close your books, put away your materials, and sing the song to end ML.

LESSON C7

Focus: Introduction to Addition.

Materials: As before.

Morning Exercises:

Through the Night [song]
Through the night, the Angels kept
Watch beside me while I slept.
Now the dark has gone away.
We are glad for this new day.

Morning Has Come [song]
Morning has come, night is away.
We rise with the sun and welcome the day.

Round and Round [song with movement]

Round and round, the Earth is turning,	*[Standing straight and tall, turn slowly in*
Turning always round to morning,	*place until you are facing forward at*
And from darkness round to light.	*the end of the song.]*

Tall Trees in the Forest [verse with movement]
[Repeat three times, varying speed of movement once the child knows the verse.]

Tall trees in the forest;	*[Standing straight and tall, lift up on toes and reach high with your arms.]*
Pine cones on the ground.	*[Slowly bend over and "touch" the pine cones.]*
Tall trees in the forest,	*[Standing straight and tall.]*
They bend, they bend, they bend.	*[Bend over slowly and touch the floor.]*

Stomping Land [traditional children's song]

I traveled far across the sea,	*[Put hand to forehead, look into distance]*
I met a man, and old was he.	*[One hand on "cane," the other on bent back]*
"Old man," I said, "Where do you live?"	*[Hands on hips.]*
And this is what he told me:	*[Index finger 'bouncing' for emphasis]*
"Follow me to stomping land,	*[Stomp]*
Stomping land, stomping land.	
All who wish to live with me,	
Follow me to stomping land.	*[In successive verses, choose an action and do it.]*

Bean Bags

Cross-patch, draw the latch,	*[motions as before]*
Sit by the fire and spin.	
Take a cup and fill it up,	
Invite the neighbors in.	

Whisky, frisky, hippety-hop!	*[Start with the bean bag held in front and toss*
Up he goes, to the tree top!	*it around your body in time with the verse.]*
Whirly, twirly, round and round,	
Down he scampers to the ground.	
Furly, curly, what a tail!	
Tall as a feather, broad as a sail.	
Where's his supper? In the shell.	
Snappy-cracky, out it fell.	

Lovely *rain*bow, **hung** so *high*,	*[Bring hands together over head and switch*
Glistening *in* the **clear**ing **sky.**	*bean bag from one hand to the other on*
	bold *syllables.*
Red and *orange,* **yel**low and *green,*	
Rainbow *colors* **bright**ly *gleam.*	*[On italic syllables, hold hands*
	out to sides, arms parallel to floor.]
Blue and *indi*go, **pur**ple so *fine,*	
Rainbow *colors* **soft**ly *shine.*	
Pots of *gold* may **there** be *found,*	
Where the *colors* **meet** the *ground.*	

Teddy Bear [Follow instructions in verse for motions.]
Teddy Bear, Teddy Bear, turn around.
Teddy Bear, Teddy Bear, touch the ground.
Teddy bear, teddy bear, reach up high,
Teddy bear, teddy bear, touch the sky,
Teddy bear, teddy bear, bend down low,
Teddy bear, teddy bear, touch your toes,
Teddy bear, teddy bear, go to bed,
Teddy bear, teddy bear, rest your head,
Teddy bear, teddy bear, turn out the lights,
Teddy bear, teddy bear, say "good night".

You Are One [by Vivian Jones-Schmidt]
[Hold up the correct number of fingers for each verse.
For 11 and 12, hold up ten and then one or two more, as needed.]
You are one and I am one,
And one is the Golden Sun.

Golden Sun and Moon make two,
And so do I and you.

Sun and Moon and Earth will be
Always together and always three.

Fall and Winter, add two more.
Spring and Summer make seasons four.

I am a star that shines so bright,
With five strong points to make my light.

The cells within each hive of bees
With six sides hold the honey sweet.

Within each rainbow in the sky
Seven colors stretch side to side.

Sparkling in the morning dew
Spider's eight legs weave a web that's new.

Nine are the candles glowing bright
That sit in the window in the dark night.

Ten fingers have I, five on each hand,
And ten strong toes that help me stand.

Eleven hours of sleep at night
Keep me happy in the sunshine bright.

Every year is twelve months long,
And each month sings a different song.

Counting with Play Ball
Count to 24, emphasizing the even numbers.
Hold the ball in your hands on the odd numbers, and bounce on the even numbers.
When your child can do this without hesitation, whisper on the odd numbers and shout on the even numbers.
Count forward from one to twelve and then back from 12 to 1.

I Am a Tall Tree [verse with movement]
I am a tall tree, reaching up to the sky.	*[Standing straight, stretch arms up.]*
When the wind blows, I lean with a sigh.	
I lean to the front,	*[Legs straight, lean to the front.]*
I lean to the back,	*[Legs straight, lean to the back.]*
I stand up straight without a crack.	*[Jump back into uprightness.]*

Straight as a Spear [verse with movement]
Straight as a spear, I stand.	*[Standing as tall and straight as possible.]*
Strength fills my legs and arms.	*[Spread legs apart, stretch arms straight out to sides.]*
Warmth fills my heart with love.	*[Cross arms over heart.]*

[Swinging your arms around to each side, jump back into upright position, arms at sides.]

The Sun with Loving Light [verse with no movement]
[Stand up straight and tall, crossing arms over the heart, with reverence.]
The Sun with loving light makes bright for me each day.
The Soul with Spirit power gives strength unto my limbs.
In sunlight shining clear, I do revere, O God,
The strength of humankind which you so graciously
Have planted in my soul,
That I with all my might
May love to work and learn.
From you stream light and strength.
To you, rise love and thanks.

I Look at My Hands [verse with gestures]

I look at my hands with my fingers fine,	*[Hold hands out in front of you.]*
And I want to be proud that they are mine.	
For deep in my heart lies a golden chest	*[Cross hands over heart.]*
With secret treasures that no one can guess	
Unless my hands do their very best.	*[Hold hands out in front of you.]*

Warm Our Hearts [song to end Morning Lesson]

Warm our hearts, O Sun, and give	*[Cross hands over heart.]*
Light that we may daily live,	
Growing as we want to be:	
True and good and strong and free.	

Lesson Content:

1. The next friend I want you to get to know is called Little Plus. Plus loves to share, and whenever she has apples or dimes or muffins, she can't wait to share them with his friends. Plus goes for walks every day, looking for beautiful stones to share with her friends.

2. When she finds a stone, she puts it in one of her pockets. She picks up stones until she has 12, but sometimes she can't remember how many stones she has put in each pocket.

3. "Now we're going to draw Little Plus. Open your book to **page 30.** First, lay four block crayons across the top of your page, widest side down. Make a pencil mark after the fourth crayon. This mark will divide the page into approximately two equal sections. Now take a stick crayon and trace the straightest line you can, in the air, from the top to the bottom of the page. Then draw it on the page.

4. With your brown and green block crayons, draw the earth from the left side of the page to the vertical line you just drew. [You will draw an "earth" using two vertical bars of color.]

5. "Choose a brown to draw her face and hands. Draw her face like I draw mine." Until your child is comfortable with drawing people, it will be a good idea for you to place your finger on your child's page in the place for the head.

6. With your green stick crayon, draw the trunk and limbs of Little Plus, arms outstretched and with hands in pockets. Draw the feet, then hair and a cap. Draw a "plus" sign on her shirt.

7. Draw a bright sun above Little Plus and swirls of blue for the sky.

8. "This is Little Plus. In our next class, she will show us how she works."

9. Close your books, put away your materials, and sing the song to end ML.

Focus: Introduction to addition algorithm.

Materials: As before.

Morning Exercises: As before.

Lesson Content:

1. Open your books to **page 30.** "What is the name of this helper? This friend is called Little Plus, and she helps us whenever we need to add numbers together." Create four work spaces on the right hand side, just as you created them on **page 29.**

2. "Take your stones out of their pouch and count them one by one."

3. "Little Plus goes for a nice walk every day to look for stones, and every day she comes home with 12 stones. She keeps the stones in her pockets while she is walking. When she gets home, she empties her pockets and counts the stones in each."

4. "One day, she empties her right pocket and finds four stones in it. She knows she will have 12 stones altogether, so how many stones will be in her left pocket?" Demonstrate counting 4 stones into a little pile or line, and then counting the remaining 8 stones, starting over at "one."

5. "Yes, 12 is the same as 4 plus 8. Now I'm going to teach you how to write this story in numbers." First, write a large 12 in the top work space. Now draw 4 large dots above the 12. "After the 12, we write this sign: =. What do we say when we see this sign?"

6. "Now we write the number for the stones in her right pocket: 4. Let's keep drawing the stones in her left pocket, with a different color crayon so we'll remember that these stones were in a different pocket." Count the four stones already drawn, and with your other color crayon, draw large dots as you continue counting from 5 - 12.

7. Now count the "new" stones you've drawn. "These are the stones in her right pocket, and these are the stones in her left pocket. The stones are in different pockets, but Little Plus reminds us that we have to show that we are putting them all together. So we use this sign: + to show that. When we see this sign, we say, 'plus.'"

8 "Altogether there are 12 stones. Let's read this arithmetic sentence together: 'Twelve equals/is the same as four plus eight.' Now you read the sentence by yourself."

9 Repeat this process with the following number stories, recording them in the work spaces on **page 30:**

   a. 12 = 3 + 9
   b. 12 = 5 + 7
   c. 12 = 10 + 2

There are of course several other combinations of addition factors for twelve, and you are free to choose any combinations you like for writing in your books. If there is time, encourage your child to discover them by counting out their stones.

10 Practice reading aloud the equations/number stories.

11 Close your books, put away your materials, and sing the song to end ML.

LESSON C9

Focus: Introducing Subtraction.

Materials: As before.

Morning Exercises: As before.

Lesson Content: Introducing Subtraction.

1. Show your child your illustrations of Little Divide and Little Plus, and make sure they know the names and stories of these two Arithmetic Friends.

2. "Today I'm going to tell you about another friend. This helper is called 'Little Minus.' Little Minus is often very sad, because he keeps losing things! Have you ever lost anything?" Discuss how it feels to lose something. "Little Minus really wants to share the things he finds with other people. When he sets out to share his 12 jewels, he counts them to make sure there are 12. But when he gets to his friend's house, he only has 7! Poor Little Minus, he is always disappointed."

3. Open your book to **page 31**. Divide the page into two sections as before. On the left hand side, draw the earth with the green and brown block crayons.

4. Now draw Little Minus. Choose a brown for his skin tone and draw his head. Then with your blue stick crayon, draw his body, in profile, bending over with sadness.

5. Draw his hair, hands, and feet, and a cap. Draw a pocket with an obvious "hole" in it, and jewels falling out. Now draw the sun and sky.

6. Close your books, put away your materials, and sing the song to end ML.

LESSON C10

Focus: Introducing the subtraction algorithm.

Materials: As before.

Morning Exercises: As before.

Lesson Content:

1. Open your books to **page 30.** "What is the name of this arithmetic helper? What is his story?"

2. "This is Little Minus, and he also loves to share. But he keeps losing things! Why does he lose things? Yes, there's a hole in his pocket. Poor Minus!"

3. "Open your pouch, and count out your stones. Make sure there are 12. Now pretend you're Little Minus. You start from home with 12 stones to share with your friend. But when you get to her house, you only have 5 stones! How many did you lose?" Some children will understand immediately that they need to move 5 stones to one side and count the remaining stones. Some children will need help figuring this out. If your child does not know what to do, demonstrate the process for them, using the stones on the desk.

4. "Yes, if you leave home with 12 stones and arrive with 5 stones, you have lost 7 stones. Now let's say that you leave with 12 stones and arrive with 8. How many did you lose?" Watch as your child works this out with their stones. "Yes, you lost 4 stones. We say, '12 minus 4 is the same as 8.'"

5. Continue with one or two more subtraction stories. Then prepare the right-hand side of **page 31** as you prepared the previous two pages. You will have four work spaces on the right hand side of your page.

6. Please note: When we write the subtraction algorithm, we start with the original quantity [12]; write the equal sign; draw a blank; and then write the number we "arrived" with. The number of stones lost is the unknown quantity. In the number language of arithmetic: 12 - __ = 8.

7 Create four subtraction stories for your child and write them in the four work spaces.

8 Practice reading aloud the four equations/ number stories.

9 Close your books, put away your materials, and sing the song to end ML.

FIRST GRADE ARITHMETIC—TEACHER'S GUIDE   95

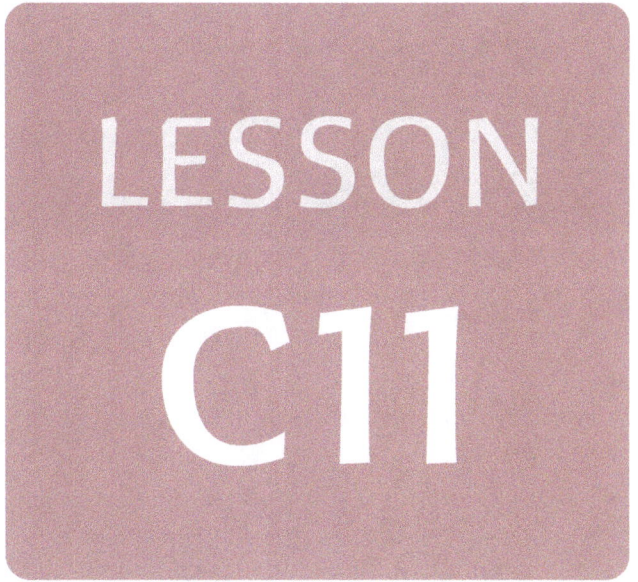

Focus: Introduction of Multiplication/

Materials: As before.

Morning Exercises: As before.

Lesson Content:

1 "Who are the arithmetic friends we've met so far?" Review Divide, Plus, and Minus.

2 "Today, we're going to meet one more helper. This friend likes to share also, and she likes to work really fast! Take your stones out of their pouch and we'll see how this friend works. Be sure you have 12."

3 Count out 12 stones. "Now, our new friend wants to give 3 stones to each of her friends. How many friends can she give 3 stones to?" Demonstrate how to count out piles of 3 stones. "How many piles do you have?"

4 Continue with this exercise, asking, "How many friends can she give 2 stones to?..... 4 stones?.....6 stones? 12 stones? One stone."

5 "Now open your book to **page 32** so we can draw this friend, whose name is 'Times.'" Set up **page 31** as before.

6 On the left hand side, draw the earth with your green and brown block crayons. "Times works so quickly, she seems to turn cartwheels instead of walking. And that's how we're going to draw her."

7 Choose a skin tone and draw a face for Times in the center of this side of the page, about one inch from the earth. Now take your yellow stick crayon and draw Times upside down: arms and legs outstretched until she forms an "X." Draw her hair, hands, and feet. [No cap, because it would probably fall off!] Draw the sign for "times" on her shirt, and point out that this is the sign we draw when we work with Times.

8 Draw a golden sun above her and a blue sky.

9 Close your books, put away your materials, and sing the song to end ML.

LESSON C12

Focus: Introducing the multiplication algorithm.

Materials: As before.

Morning Exercises: As before.

Lesson Content:

1. "Count out your stones to make sure you have 12. Now open your book to **page 32**."

2. Set up the right hand side of the page as before.

3. "Now I'm going to teach you how to write the stories we tell about Little Times. Across the top of this work space, draw 12 stones."

4. "Times wants to give each of her friends 3 stones. How many friends can she share 3 stones with? We circle 3 stones as many times as we can."

5. "Now, we started with 12 stones, so we write **12** here. What is our sign that means 'is the same as?'" Draw the **equal** sign after the **12**.

6. "We don't know how many times we'll be able to share 3 stones, so now we have to draw a blank space. Then we draw the sign that means **times**, and we write the **3**. How many times were we able to share 3 stones? Write the **4** in the blank."

7. Repeat this process with three more combinations of numbers, choosing from:

   12 = ___ x 2   12 = ___ x 4   12 = ___ x 1
   12 = ___ x 6   12 = ___ x 12

8. Practice reading aloud all four equations/ number stories.

9. Close your books, put away your materials, and sing the song to end ML.

MAIN LESSON BLOCK D

Extension of Operations

THIS BLOCK BEGINS WITH A LESSON ON ODD AND EVEN NUMBERS. What makes a number odd or even? Numbers that have "partners" are even; those that do not have partners are "odd." Make sure your paintings of numbers 1 - 12 are displayed, as they are a helpful visual aide in explaining these concepts. The blue numbers are sad because they don't have partners; the red numbers are happy because they do. Your child might very well notice that all of the even numbers are multiples of two!

Next, we'll spend a little time working with the numbers 1 - 100. The initial illustration is that of gnomes gathering jewels deep underground and placing them in rows of ten on ten shelves. On the facing page, the jewels become numbers. Spend some time talking about the patterns the numbers make, vertically and horizontally. Being able to find and understand patterns is an important mathematical skill.

In Block C, you introduced your child to the four basic operations of arithmetic: addition, subtraction, multiplication, and division. In that block, we worked with the numbers 1 - 12. In Block D, we return to these basic operations, this time working with numbers 1 - 24. We also continue to skip count, as our goal by the end of First Grade is to comfortably count to 100 by ones and to skip count:

By twos to 24
By threes to 36
By fives to 100
By tens to 100

Many children also quickly catch on to skip counting by 11s at least to 99. When you skip count with your child, just say the numbers: 2, 4, 6, 8, etc. In the same session, review the multiplication tables: $2 = 2 \times 1$; $4 = 2 \times 2$; $6 = 2 \times 3$; etc., until you reach $24 = 2 \times 12$. In a similar way, review the tables for 3, 5, and 10.

Although some adults might have uncomfortable memories about learning their multiplication tables and addition facts, research emphasizes that it is important to develop these skills. Why? When children can automatically recall these basics, they are free to think about how to approach particular number questions rather than spending their time–and brain power–trying to figure out, for example, the answer to 5+4 or 7 x 3. Moreover, having these "facts" at their fingertips allows them to manipulate numbers to solve questions in creative ways. If I know that 9x8=72, I can quickly figure out that 9x16=144.

Woven into this material are lessons that present, in story form, the operations of arithmetic: how numbers can be taken apart and put back together. In this block, division, addition, subtraction, and multiplication are all taught with the story of Farmer Jo/e. [I chose this name because the farmer can be either gender; of course you can choose the name and the gender as you wish. Please treat this matter-of-factly, as the point is not the farmer's name or gender, but rather the arithmetic.] The farmer collects eggs and gives them to neighbors in the first instance. When we review addition, the farmer needs 24 eggs but can only find x number of eggs. How many more are needed? When we talk about subtraction, drama enters our story when a stealthy rat breaks into the henhouse and eats the eggs!

As in every subject, it is important to approach learning the basic multiplication and arithmetic combinations with joy and enthusiasm. Smile! Laugh when your play ball flies out of your fingers; then jump right back to wherever you left off. Alternate the tables and addition or subtraction facts with other activities. Find opportunities outside of class to practice with the playball and jump rope. Traditional jump rope rhymes can make practice easy and fun.

Focus: Odd and even numbers.

Materials:
- Bean bag
- Copper rod
- ML Book #2
- Block and stick crayons
- Large colored pencils
- 24 Counting stones

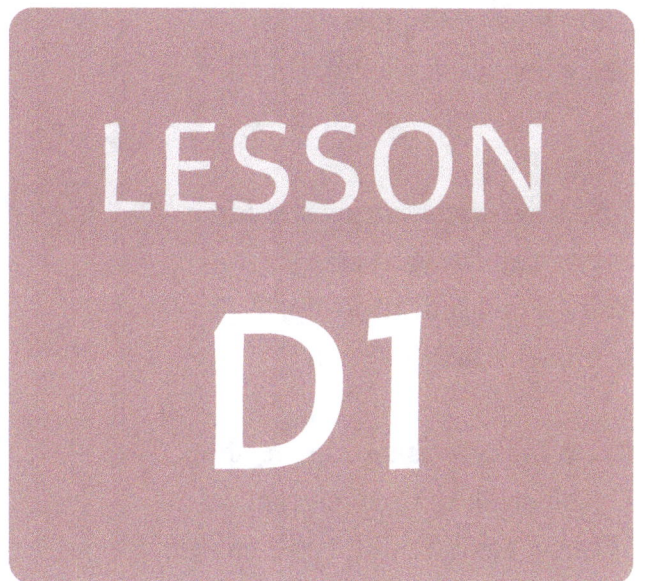

Morning Exercises:

Through the Night [song]
Through the night, the Angels kept
Watch beside me while I slept.
Now the dark has gone away.
We are glad for this new day.

Morning Has Come [song]
Morning has come, night is away.
We rise with the sun and welcome the day.

Round and Round [song with movement]

Round and round, the Earth is turning,	*[Standing straight and tall, turn slowly in place until you are facing forward at the end of the song.]*
Turning always round to morning, And from darkness round to light.	

Tall Trees in the Forest [verse with movement]
[Repeat three times, varying speed of movement once the child knows the verse.]

Tall trees in the forest;	*[Standing straight and tall, lift up on toes and reach high with your arms.]*
Pine cones on the ground.	*[Slowly bend over and "touch" the pine cones.]*

Tall trees in the forest, [Standing straight and tall.]
They bend, they bend, they bend. [Bend over slowly and touch the floor.]

Bean Bags
#1. Cross-patch, draw the latch, [motions as before]
Sit by the fire and spin.
Take a cup and fill it up,
Invite the neighbors in.

#2. Whisky, frisky, hippety-hop! *[Start with the bean bag held in front and*
Up he goes, to the tree top! *toss it around your body in time with the*
Whirly, twirly, round and round, *verse.]*
Down he scampers to the ground.
Furly, curly, what a tail!
Tall as a feather, broad as a sail.
Where's his supper? In the shell.
Snappy-cracky, out it fell.

#3. **Lov**ely *rainbow,* **hung** so *high,* *[Bring hands together over head and switch bean*
Glistening *in* the **clear**ing *sky.* *bag from one hand to the other on* **bold** *syllables.*
Red and *orange,* **yel**low and *green,* *On italic syllables, hold hands out to sides,*
Rainbow *colors* **bright**ly *gleam.* *arms parallel to floor.]*
Blue and *in*digo, **purp**le so *fine,*
Rainbow *colors* **soft**ly *shine.*
Pots of *gold* may **there** be *found,*
Where the *colors* **meet** the *ground.*

#4. **Wa**ter, **wa**ter, water **fall.** *[Hold bean bag in right hand behind head.*
Wher**ev**er I **am,** I **hear** your **call.** *On* **bold** *syllables, drop bean bag into your*
You are **love**ly, **spark**ling, **tall.** *left hand, which is held at mid-back,*
You **make** of **wa**ter, **one** strong **wall.** *cupped to catch the bean bag.]*

#5. **Jack** be **nim**ble, **Jack** be **quick!** *[Hold bean bag in the right hand. On the* **bold**
Jack jump **ov**er the **can**dle **stick!** *syllables, switch hands* **under** *your left knee.*
  Then, with the bean bag in the left hand, switch
  to the right hand under the right knee.
  Continue the pattern, varying the Speed.]

Copper Rod Exercises:

These introductory copper rod exercises "wake up" the hand and arm muscles needed for writing and other skills.

#1: Hold the rod in both hands, palms up. Reciting "Wee Willie Winkie," gently roll the rod back and forth, catching it with the thumb so that it does not progress beyond the hand.
Wee Willie Winkie runs through the town,
Upstairs and downstairs in his nightgown.
Rapping at the windows, crying through the locks:
"Are the children all in bed? For now it's eight o'clock!"

#2: This time, recite "Hickory Dickory Dock" as you allow the rod to gently roll all the way to your elbows and back. *[Feel free to make up your own verses, as I have here!]*
Hickory Dickory Dock, the mouse ran up the clock.
The clock struck one, the mouse ran down,
Hickory Dickory Dock.
Hickory Dickory Dock, the mouse ran up the clock.
The clock struck two, the mouse sneezed, "Achoo!"
Hickory Dickory Dock.
Hickory Dickory Dock, the mouse ran up the clock.
The clock struck three, the mouse called, "Whee!"
Hickory Dickory Dock.

#3. Now sing "Row, Row, Row Your Boat" while allowing the rod to roll all the way up to your shoulders and back to your fingertips.
Row, row, row your boat gently down the stream.
Merrily, merrily, merrily, merrily: Life is but a dream.

#4. Humpty Dumpty
[Start with the copper rod held horizontally at waist height in both hands, palms down. With the first two lines, raise the rod over your head; with the second two lines, lower the rod as far as you can without bending over.]
Humpty **Dump**ty
Sat on a **wall.**
Humpty **Dump**ty
Had a great **fall.**

[Start these next two lines holding the rod horizontally at the waist. Without shifting your hands from the previous position, first move the rod to the vertical with the right hand on top, then the left.]
All the King's **horses**
And **all** the King's **men**
[The rod is once again at waist height.
On the first line, lift it over your head; on the second line, lower it as before.]
Could **not** put **Humpty**
To**geth**er a**gain.**

#5. Now place the rod carefully on top of your head and, walking slowly, recite this verse while following the directions for additional movement. This is a challenge!
The Stork
I lift my leg, I stretch my leg,
I plant it, firm and light.
I lift again, and stretch again,
My pace, exactly right.
With care I go, so grand and slow–
I move just like a stork.
My eye is bright, my head upright,
And pride is in my walk.

#6. Hold the copper rod with both hands, palms down, in front of you. Reciting this poem, lift one finger at a time as indicated. Start with your right hand, then lift the fingers of the left hand, then move the fingers of both hands simultaneously. As you become accustomed to the pattern, vary the speed.

Peter	*[index finger]*
Peter	*[middle finger]*
Pumpkin	*[ring finger]*
Eater	*[little finger]*
Had a	*[little finger]*
Wife but	*[ring finger]*
Couldn't	*[middle finger]*
Keep her.	*[index finger]*
He put her	*[index finger]*
In a	*[middle finger]*
Pumpkin	*[ring finger]*

Shell   *[little finger]*
And there   *[little finger]*
He kept her   *[ring finger]*
Very   *[middle finger]*
Well.   *[index finger]*

Straight as a Spear [verse with movement]

Straight as a spear, I stand.   *[Standing as tall and straight as possible.]*
Strength fills my legs and arms.   *[Spread legs apart, stretch arms straight out to sides.]*
Warmth fills my heart with love.   *[Cross arms over heart.]*
[Swinging your arms around to each side, jump back into an upright position, arms at sides.]

The Sun with Loving Light [verse with no movement]

[Stand up straight and tall, crossing arms over the heart, with reverence.]
The Sun with loving light makes bright for me each day.
The Soul with Spirit power gives strength unto my limbs.
In sunlight shining clear, I do revere, O God,
The strength of humankind which you so graciously
Have planted in my soul,
That I with all my might
May love to work and learn.
From you stream light and strength.
To you, rise love and thanks.

I Look at My Hands [verse with gestures]

I look at my hands with my fingers fine,   *[Hold hands out in front of you.]*
And I want to be proud that they are mine.
For deep in my heart lies a golden chest   *[Cross hands over heart.]*
With secret treasures that no one can guess
Unless my hands do their very best.   *[Hold hands out in front of you.]*

Warm Our Hearts [song to end Morning Lesson]

Warm our hearts, O Sun, and give   *[Cross hands over heart.]*
Light that we may daily live,
Growing as we want to be:
True and good and strong and free.

Lesson Content:

1. Count out all 24 of the counting stones, slowly and carefully. Your child should place a finger on each stone when counting. Make sure your child is counting one stone at a time. [If your child moves more than one stone with each number, this means that they do not yet understand "one-to-one correspondence," and they will need much practice in counting "one number, one stone."]

2. Instruct your child to place the stones in pairs, making a column of two stones. [There should be twelve pairs.]

3. Now ask your child to count the stones again, starting at the top left with "one" and counting "two" as the stone on its right. Do this several times, varying the speed of the counting.

4. Explain to your child that, when numbers have a partner, they are called "even numbers." Ask your child to name the even numbers in the stones before them.

5. Now explain to your child that if a number does not have a partner, it is called an "odd number." Ask your child to name the odd numbers in the stones before them.

6. An engaging activity at this point is to look around your house and ask your child questions like:
   a. Do we have an even number of chairs or an odd number of chairs?
   b. Are there an even number or an odd number of apples in the bowl?
   c. Are there an even number or an odd number of windows in this room [or in our house]?
   d. Are there an even number or an odd number of eggs in our refrigerator?

7. Feel free to move around the house and/or yard while asking the questions. When you have engaged in this activity for several minutes, take out your ML #2 books. Turn to **page 33.**

8. With two block crayons on the longer side, draw alternating bars of color on your page.

9. Write the numbers 1 - 24 in these combinations: 1-6, 7-12, 13-18, 19-24.

10. Now take a different color of stick crayon and explain to your child that you are going to put the numbers in pairs on your page. Draw a bracket under each two numbers: 1-2, 3-4, 5-6, 7-8, etc.

11. Ask your child to name the even numbers on the page; then ask your child to name the odd numbers.

12. Close your books, put away your materials, and stand for the closing verse.

Focus: Counting to 100.

Materials:
- ML Book #2
- Block and stick crayons
- Bean bag
- Copper rod
- Adult drawing of the lesson

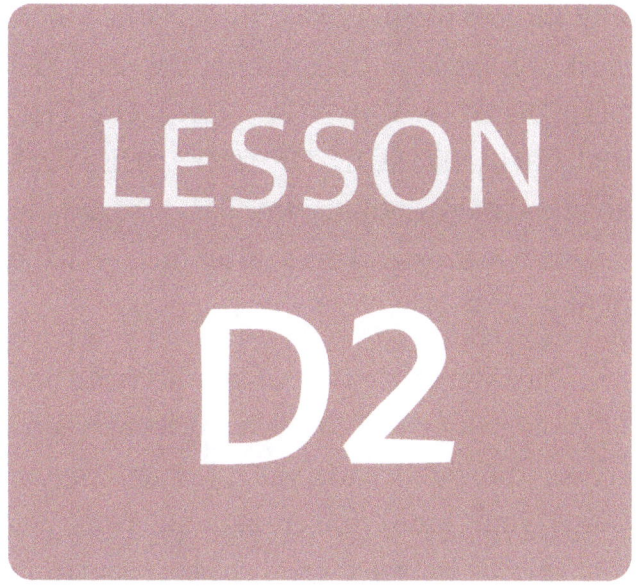

Lesson Content:

1 This lesson is a visual imagination of the first 100 numbers. Say to your child: "We are so fortunate to have so many beautiful jewels in our world. There are jewels of every color we can think of and even colors we don't think of. Think really hard. Do you know the names and colors of any jewels?" [Some possibilities are: amethyst [purple], sapphire [blue], emerald [green], topaz [yellow], amber [orange], ruby [red], onyx [black], pearl [white]. After class, finding pictures of minerals in library books could be a nice extension of this lesson.

2 "All of these jewels can be found right here, in different places under our Earth. I like to imagine that they are all in one big cavern under the ground, and that they are cared for by small beings that some people call gnomes. Today, we're going to draw an illustration of what that place might look like." At this point, show your child your drawing. Plan to draw the illustration again at the same time your child is drawing theirs.

3 "Open your book to **page 34.** Across the bottom of your page, draw two bars of brown and purple with the widest side of your crayon, all the way across the page from left to right. This is the floor of the cavern. Now make a little mark in the middle of your page with your yellow stick crayon. This will help us know where to start the shelves for our jewels."

4 Place a block crayon against the spiral binding of your page, long side next to the binding. Make a little mark. This is where you'll draw the bottom shelf.

5 Use four block crayons stacked on top of one another to measure from the top of your page, and make another mark at the bottom of that stack. This is about how tall your shelves will be.

6 With a brown stick crayon, draw a vertical line, from the measured mark you made

with the four block crayons, through the yellow mark you made in the center of the page. DO NOT DRAW A VERTICAL LINE ON THE RIGHT-HAND SIDE OF THE SHELVES. You need to leave this space open, because you might have to extend the shelves further to the right.

7 Starting at the top of the vertical line you just drew, draw a horizontal line across the page with your yellow crayon. Using your index finger as a measure for the space between the lines, continue drawing horizontal lines until you have drawn eleven. [You are drawing with your yellow crayon in case you need to adjust a line.] When your shelves have been drawn so that there are eleven, between the top two lines, draw the "jewels"--small circles of color in this order: magenta, red, vermillion, orange, golden, yellow, yellow-green, green, blue, purple. [If your crayon set does not have a yellow-green, make it by drawing a circle in yellow and lightly shading it with green.]

8 Now you know how long your shelf has to be to accommodate ten "jewels." With your brown stick crayon, draw a vertical line on the right side of the shelves. Then draw all eleven shelves, covering up the yellow "helper" lines with the brown.

9 Continue to draw the jewels, in that order, until all ten shelves are filled.

10 When all 100 jewels have been drawn, close your books, put away your materials, and say the closing verse.

FIRST GRADE ARITHMETIC—TEACHER'S GUIDE

Focus: counting to 100, completing the illustration

Materials:
- ML Book #2
- Block and stick crayons
- Bean bag
- Copper rod
- Blank 100 chart [see appendix]

Morning Exercises: As before.

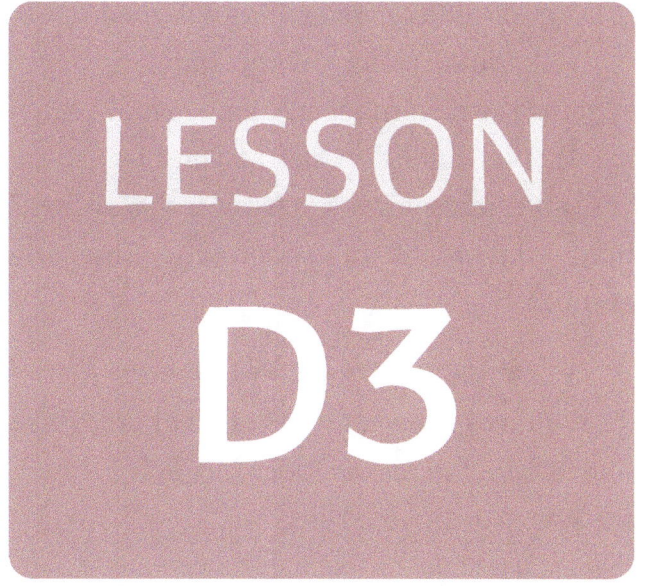

LESSON D3

Lesson Content:

1. Open your book to **page 34**. "Yesterday we drew this illustration. There are a lot of jewels here! Let's count them to see how many we have drawn." Count the 100 jewels, from left to right and top to bottom. Make sure your child places a finger on each jewel as it is counted.

2. "100 jewels! That is a lot! If there are small beings taking care of these jewels, they have a lot of work to do. Let's draw them doing some of that work."

3. With your stick crayons, draw "gnomes" in a variety of work poses: polishing the jewels, moving jewels in a wheelbarrow, holding a ladder, etc. Samples are shown on page 106.

4. Now, with brown and green block crayons, add a curving line above the shelves to represent the surface of the earth. You can also add flowers [with roots] and a tree [with roots].

5. When your child has finished this drawing, collect their book and put it away. "We've drawn 100 jewels, and we've counted to 100. Now we're going to write the numbers from 1 to 100." Give your child a blank 100 chart and a graphite pencil.

6. On your chalkboard or on a very large piece of paper, you'll be carefully drawing each number for your child to copy. Set your pace to that of your child. Some children will need to see one number at a time, while others are not overwhelmed by two or three.

7 Your first line of numbers will read 1 - 10. Ask your child, "What happens now? What is the next number? Where shall I draw it?" If your child does not respond by saying "11" and "in the first square of the next line," help your child to understand this. Be encouraging.

8 Continue to draw the numbers in the squares. You may only be able to fill in 3 or 4 lines in this lesson, but you'll finish the chart in the next lesson.

9 Close your books, put away your materials, and stand for the closing verse.

Focus: Finish writing the numbers in the 100 chart.

Materials:
- ML Book #2
- Block and stick crayons
- Bean bag
- Copper rod
- Graphite pencil
- 100 chart

Morning Exercises: As before.

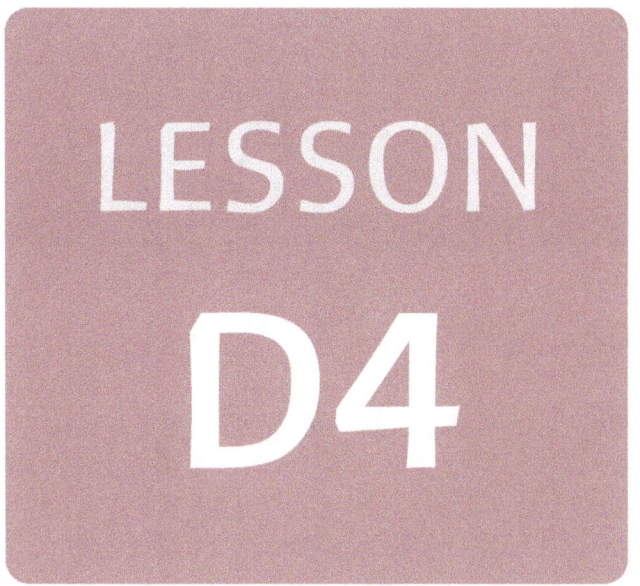

Lesson Content:

1. Take out the 100 chart you started yesterday. Finish filling in the numbers with the graphite pencil.

2. Now ask your child, "What do you notice about the numbers?" Your child will probably notice that every line starts with a 1; that the second number in each line follows the pattern 1-2-3....0; that all of the numbers in the last column end in zero; that the numbers in the center of the chart end in 5; that every other number is even, and the others are odd; etc. Encourage your child in finding patterns. This is fun and so important for foundational mathematics, which is all about patterns!

3. When your child has completed the chart, tape it into their book on **page 35.** Now you can draw little figures representing the four operations, one at each corner. This is a good time to remind your child of these operations. "Let's draw our arithmetic helpers at the corners of our 100 chart. Which one shall we put here [upper left]?" When your child responds, ask your child what they remember about the work this friend helps with. If it's addition, what does that mean?

4. Repeat for the remaining three helpers.

5. Tell your child: "In our next lesson, we're going to meet a friend called Farmer Jo/e." [I chose this name because it can be male or female. You are of course free to choose your own name.] "Farmer Jo/e lives on a farm and has a lot of chickens. What do chickens do? Yes! They lay eggs, sometimes every day. In our next lesson, we'll learn about what happens to those eggs."

6. Close your books, put away your materials, and stand for the closing verse.

LESSON D5

Focus: Division review.

Materials:
- Bean bag
- Copper rod
- *Modeling beeswax

Morning Exercises: As before.

Lesson Content:

1. Return to the desk and sit. *Take out one stick of beeswax for modeling "baskets."

2. Break the beeswax stick in half. You'll be using both halves, one at a time, in this lesson. Place a half stick in your hands or under your arm to warm it. While waiting, talk about the beeswax: how the little bees work so hard to make the wax in their hives. If possible, buy honey still in the comb. Show your child the little cells filled with honey and count the perfect 6 sides of each cell. You are helping your child to understand how much work goes into a finished product–beeswax for modeling and for crayons–that we often take for granted. We are grateful to the bees, the flowers, the sun and all of nature for what we have been given.

3. When the stick is warm and pliable, take it into your hands and put another half stick in your "oven." Split your half stick into six equal parts. Model each one into a small cup-shaped basket.

4. Split the other half stick into six equal parts also, and model six more "baskets."

5. Using a small piece of a different color of beeswax, roll 24 "eggs."

6. Place your baskets and eggs in a safe place.

7. Now open your books to **page 36.** Draw Farmer Jo/e with baskets of eggs around. Decide which gender you will use for "Farmer Jo/e," and draw the farmer accordingly. I anointed my farmer as a female, but you are free to choose. "Farmer Jo takes good care of her hens, and they lay many eggs. Every day she collects the eggs in an egg basket. Then she decides which neighbors she will give the eggs to. Each of her neighbors meets her at their doors with their own baskets."

8. Stand to say the closing verse.

It will be easier to manage the subsequent activities if you choose one type of color, dark or light, for the nests and the opposite, light or dark, for the eggs.

LESSON D6

Focus: Division.

Materials:
- ML Book #2
- Stick and block crayons
- Large colored pencils
- "Baskets" and "eggs" from the previous lesson
- Bean bag
- Copper rod

Morning Exercises: As before.

Lesson Content:

1. Open your books to **page 37.** Divide the page into 8 sections as you have in previous lessons. Count out your eggs and make sure there are 24. "On Monday, Farmer Jo chooses four neighbors to receive her eggs. How many eggs will she give to each neighbor?" Instruct your child to determine the answer to this question by using their baskets and eggs.

2. When your child has responded with the correct answer, illustrate that question in the upper left hand section of your pages: 24 divided by 4 = 6.

3. Continue to work with the story of Farmer Jo and her eggs. Each time, as the correct response is made, illustrate the question in a section of your pages.

   24 divided by 2 = 1
   24 divided by 6 = 4
   24 divided by 3 = 8
   24 divided by 8 = 3
   24 divided by 12 = 2

As you work with the questions, ask your child to assist in writing the responses. What do we write first? Which sign do we use after the 24? What does this sign [=] mean?

4. When you have completed all 8 sections of your page, close your books, put away your materials, and stand for the closing verse.

First Grade Arithmetic—Teacher's Guide

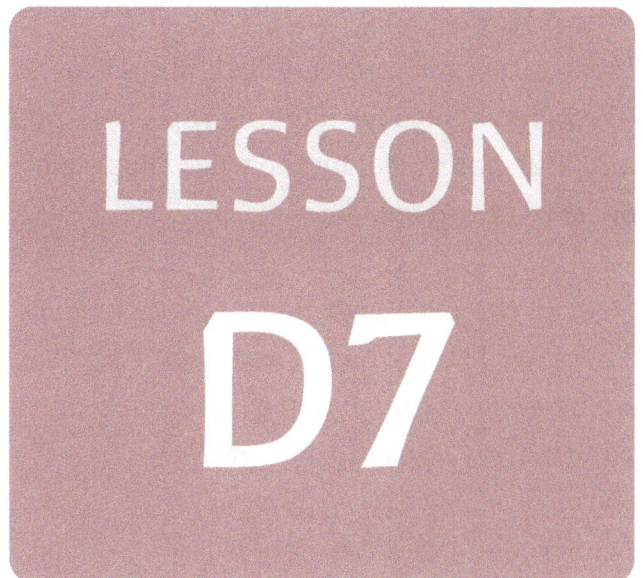

LESSON D7

Focus: Subtraction review.

Materials:
- Play ball
- Jumprope
- Copper rod
- Beeswax baskets and eggs
- Block and stick crayons
- ML Book #2

Morning Exercises:

Through the Night [song]
Through the night, the Angels kept
Watch beside me while I slept.
Now the dark has gone away.
We are glad for this new day.

Morning Has Come [song]
Morning has come, night is away.
We rise with the sun and welcome the day.

Round and Round [song with movement]

Round and round, the Earth is turning,	*[Standing straight and tall, turn slowly in*
Turning always round to morning,	*place until you are facing forward at the*
And from darkness round to light.	*end of the song.]*

Tall Trees in the Forest [verse with movement]
[Repeat three times, varying speed of movement once the child knows the verse.]

Tall trees in the forest;	*[Standing straight and tall, lift up on toes and reach high with your arms.]*
Pine cones on the ground.	*[Slowly bend over and "touch" the pine cones.]*
Tall trees in the forest,	*[Standing straight and tall.]*
They bend, they bend, they bend.	*[Bend over slowly and touch the floor.]*

Jump Rope

Ideally, by now your child can jump rope independently–and you can, too! Here are some counting rhymes for jumping rope. You might also know of others. Choose two or three and alternate them by the week. For each rhyme, you can choose to practice counting by ones, twos, threes, etc.

Over in the meadow where the green grass grows,
There sat [child's name] as sweet as a rose.
Along came a frog and kissed her/him on the nose.
How many kisses did s/he get?

Cinderella, dressed in yellow
Went downstairs to kiss her fellow.
How many kisses did he get?

1-2-3-4-5-6-7
All good children go to Heaven.
7-6-5-4-3-2-1
Jumping rope sure is fun!

Bubble gum, bubblegum in the dish,
How many pieces do you wish?   [traditional]

Hickety pickety pop!
How many times before I stop?   [Ellen Mason]

Wash the dishes, dry the dishes,
Have a cup of tea.
Don't forget the sugar–
1, 2, 3, etc.   [Erin Pollard]

Play Ball

Choose a multiplication table or addition set to practice, rhythmically, with your ball.

Examples:

Three	[bounce ball]
Is	[catch ball]
Three times	[bounce ball]
One.	[catch ball]

[Continue with: 6 is three times two, nine is three times three, etc., up to 36.]

Two	[bounce ball]
Is	[catch ball]
One plus	[bounce ball]
One.	[catch ball]
Three	[bounce ball]
Is	[catch ball]
One plus	[bounce ball]
Two.	[catch ball]

[Continue with: 4 is one plus three, 5 is one plus four, etc. You can also reverse the numbers, as in: 4 is three plus one, 5 is four plus one, etc. Play with it.]

Count backwards from 12, bouncing your ball on each number and pausing as you catch it.

Copper Rod Exercises

#4. Humpty Dumpty

[Start with the copper rod held horizontally at waist height in both hands, palms down. With the first two lines, raise the rod over your head; with the second two lines, lower the rod as far as you can without bending over.]

Humpty **Dump**ty

Sat on a **wall.**

Humpty **Dump**ty

Had a great **fall.**

[Start these next two lines holding the rod horizontally at the waist. Without shifting your hands from the previous position, first move the rod to the vertical with the right hand on top, then the left.]

All the King's **horses**

And **all** the King's **men**

[The rod is once again at waist height. On the first line, lift it over your head; on the second line, lower it as before.]

Could **not** put **Hump**ty

To**ge**ther a**gain.**

#5. Now place the rod carefully on top of your head and, walking slowly, recite this verse while following the directions for additional movement. This is a challenge!

The Stork

I lift my leg, I stretch my leg,
I plant it, firm and light.
I lift again, and stretch again,
My pace, exactly right.
With care I go, so grand and slow–
I move just like a stork.
My eye is bright, my head upright,
And pride is in my walk.

#6. Hold the copper rod with both hands, palms down, in front of you. Reciting this poem, lift one finger at a time as indicated. Start with your right hand, then lift the fingers of the left hand, then move the fingers of both hands simultaneously. As you become accustomed to the pattern, vary the speed.

Peter	*[index finger]*
Peter	*[middle finger]*
Pumpkin	*[ring finger]*
Eater	*[little finger]*
Had a	*[little finger]*
Wife but	*[ring finger]*
Couldn't	*[middle finger]*
Keep her.	*[index finger]*
He put her	*[index finger]*
In a	*[middle finger]*
Pumpkin	*[ring finger]*
Shell	*[little finger]*
And there	*[little finger]*
He kept her	*[ring finger]*
Very	*[middle finger]*
Well.	*[index finger]*

Straight as a Spear [verse with movement]

Straight as a spear, I stand.	*[Standing as tall and straight as possible.]*
Strength fills my legs and arms.	*[Spread legs apart, stretch arms straight out to sides.]*
Warmth fills my heart with love.	*[Cross arms over heart.]*

[Swinging your arms around to each side, jump back into an upright position, arms at sides.]

The Sun with Loving Light [verse with no movement]

[Stand up straight and tall, crossing arms over the heart, with reverence.]

The Sun with loving light makes bright for me each day.
The Soul with Spirit power gives strength unto my limbs.
In sunlight shining clear, I do revere, O God,
The strength of humankind which you so graciously
Have planted in my soul,
That I with all my might
May love to work and learn.
From you stream light and strength.
To you, rise love and thanks.

I Look at My Hands [verse with gestures]

I look at my hands with my fingers fine,	*[Hold hands out in front of you.]*
And I want to be proud that they are mine.	
For deep in my heart lies a golden chest	*[Cross hands over heart.]*
With secret treasures that no one can guess	
Unless my hands do their very best.	*[Hold hands out in front of you.]*

Warm Our Hearts [song to end Morning Lesson]

Warm our hearts, O Sun, and give	*[Cross hands over heart.]*
Light that we may daily live,	
Growing as we want to be:	
True and good and strong and free.	

Lesson Content:

1. "Farmer Jo/e has a big problem. She takes very good care of her hens, but she can't sleep in the hen house. And it seems like every night, a rat sneaks into the hen house and eats some of her eggs!" Take out all of your beeswax eggs.

2. "One morning, Farmer Jo/e goes into the hen house. She looks in a nest and sees where there were 14 eggs–but one egg is gone! How many eggs did that rascally rat eat?" Assemble the 14 eggs and take away one. "So fourteen minus one leaves thirteen."

3. "On another morning, she sees that her hens had laid 17 eggs, but now there are only three. How many eggs did that rascally rat leave behind?" Assemble 17 eggs and take away eggs until there are only 3 in the pile. Count the larger pile. "So seventeen minus fourteen leaves three."

4. Continue on with this story until you have told several number stories from this list. As you tell the next story, instruct your child to count out the number of eggs that were laid. They then separate the pile, leaving the number that the rat did not eat in place. How many are in the second pile?

| 20-6 | 22-5 | 18-8 | 15-10 | etc. |
| 19-2 | 21-7 | 11-4 | 14-6 | etc. |

Always summarize the calculation, as in "Twelve minus three leaves seven."

5. Open your books to **page 38.** On this page, draw Farmer Jo/e's henhouse with several hens on their nests and a rat in one corner.

6. Close your books, put away your materials, and stand for the closing verse.

First Grade Arithmetic—Teacher's Guide

Focus: Subtraction.

Materials:
- Play ball
- Jumprope
- Copper rod
- Beeswax baskets and eggs
- Block and stick crayons
- Number path

Morning Exercises: As before.

Lesson Content:

1 Review the story from the previous lesson: "In our last class, we learned that Farmer Jo/e has a problem. What is that problem? Farmer Jo/e is pretty sad! If she doesn't have all of her eggs, she doesn't have so many to share!"

2 Open your books to **page 39.** Divide the page into 8 sections, as before.

3 In the first section, upper left on the page, draw 18 eggs across the top. "Farmer Joe had 18 eggs. But the rat ate three. How many are left for her to share?"

4 Draw a line through the last three eggs in the line. Your child should respond, "There are 15 eggs left."

5 Now write the "arithmetic sentence" under the eggs, saying it out loud as you write: 18 - 3 = 15. Remind your child that the symbol [-] is read as "minus" and that the symbol [=] is read as "is the same as" or "equals."

6 Complete the remaining seven boxes with subtraction exercises of your choice.

7 Now take out your number paths [a large one for you and a small one for your child]. Say: "This is called a number path. What do you notice about it? Can you think of anything you could do with it?"

8 Demonstrate how to use the number path, using the exercises you have just recorded in your books.

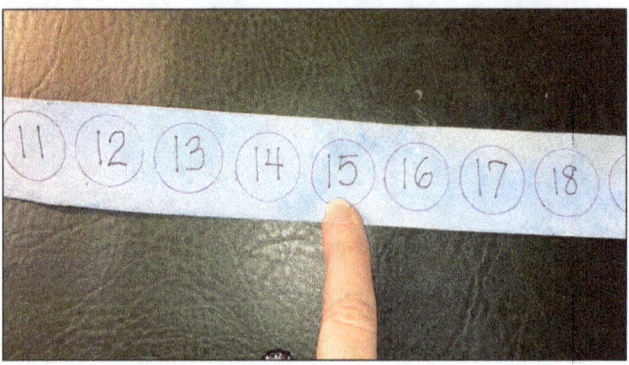

a. 18 - 3 = 15. Place your finger on the 18. Count back three numbers from the 18, placing your finger on each: 17, 16, 15. Your finger has landed on the correct answer.

b. Continue to use the number path to illustrate the remaining 7 exercises in your books. Explain that the number path is a tool that we can use in several ways. A tool helps us do our work more easily.

9. Close your books, put away your materials, and stand for the closing verse.

LESSON D9

Focus: Reviewing addition.

Materials:
- Play ball
- Jumprope
- Copper rod
- Six beeswax baskets and 24 eggs
- ML Book #2
- Stick and block crayons

Morning Exercises: As before.

Lesson Content:

1. "You know Farmer Jo/e loves to share her eggs with her neighbors. When the rat was not eating them, it was pretty easy for her to give everyone the same number of eggs, because all of her hens laid the same number of eggs! Now that the rat is busy in the hen house every night, Farmer Jo/e has to put together eggs from different nests so she can make sure that she gives each neighbor family the same number of eggs."

2. Take out the six baskets and the 24 eggs. Randomly distribute the eggs among the baskets.

3. "Farmer Jo/e has decided that, today, she wants to give each neighboring family 16 eggs. This basket has 8 eggs in it. How many more does she need to add in order to have 16?" Demonstrate for your child how to start counting at the 8 eggs in that nest and "count one more" until they reach 16. "How many did you add?" "So 16 = 8 + 8." [Remind your child that [+] is the symbol that is read as "plus."

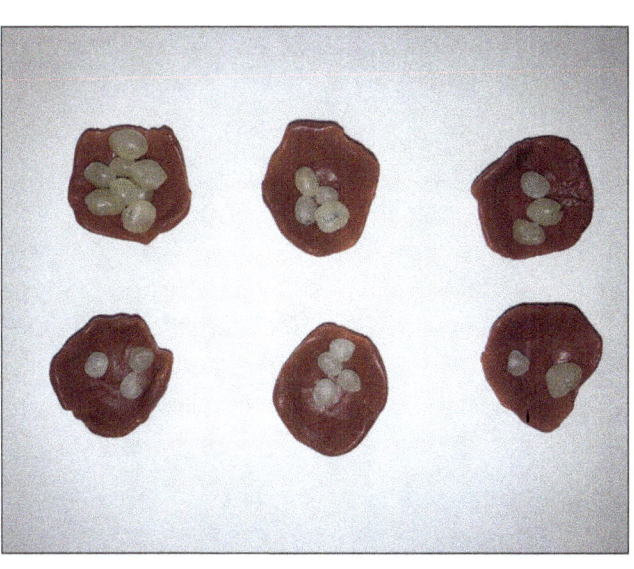

4 Here you can continue with 16 as the sum, changing the addends:

a. 16 = 12 + 4   16 = 7 + 9   16 = 10 + 6   etc.

5 "On another day, Farmer Jo/e decides that she will give each family 18 eggs. This basket has 11 eggs in it. How many more does she need to add in order to have 18 eggs?" continue as before. Again, continue with 18 as the sum and change the addends. Review the equation each time, as in, "So we know that 18 is the same as 11 + 7."

6 "What if Farmer Jo/e decides to give each family 24 eggs? If she has 12 in her basket, how many more will she need in order to give away 24?" Continue as before, remembering to restate the equation each time.

7 Now turn to **page 40** in your books. Draw a picture of Farmer Jo/e with several hens and a basket of eggs in each hand.

8 Close your books, put away your materials, and stand for the closing verse.

Focus: Addition.

Materials:
- Play ball
- Jumprope
- Copper rod
- ML Book #2
- Stick and block crayons
- Number path
- 6 beeswax baskets and 24 eggs

Morning Exercises: As before.

LESSON D10

Lesson Content:

1. Review the previous lesson. "What challenge does Farmer Jo/e face when she wants to give each neighbor family the same number of eggs?"

2. Open your books to **page 41.** Divide the page into 8 sections, as before.

3. In the upper left hand section, write the number 15. Across the top of the section, draw 6 eggs with one color of stick crayon. "Farmer Joe wants to give one neighbor family 15 eggs. She has 6 eggs in this nest. How many more does she need in order to give them 15?"

4. Using your **number path,** demonstrate for your child how to determine this. Start with your index finger on the 6. Now, count ONE and place your index finger on each number until you get to 15: 1, 2, 3, 4, 5, 6, 7, 8, 9. At "9," you should be at 15. Now say, "15 = 6 + 9." Remind your child that [+] is the symbol that we read as "plus."

5. Continue with examples until you have filled all 8 sections. For the first 2-3 examples, demonstrate for your child how to use the number path. Then expect your child to continue independently. Vary the sums, being sure to use numbers between 12 and 24.

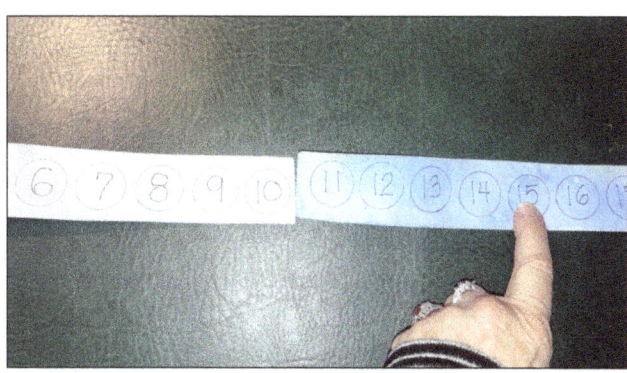

6 Close your books, put away your materials, and stand for the closing verse.

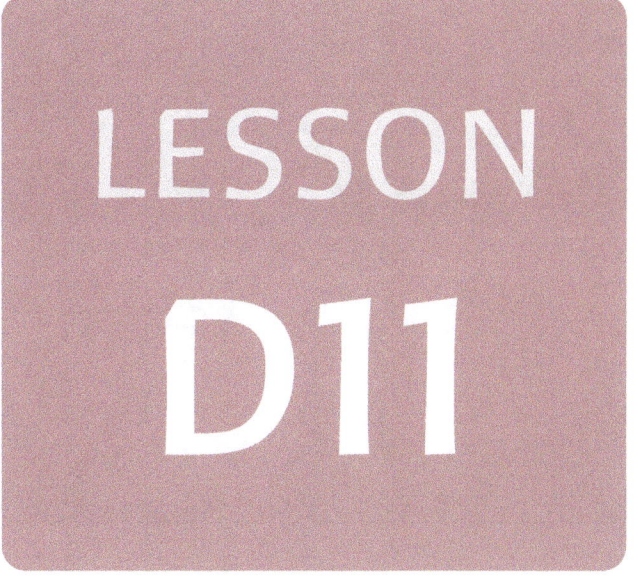

LESSON D11

Focus: Reviewing Multiplication.

Materials:
- Play ball
- Jumprope
- Copper rod
- ML Book #2
- Stick and block crayons
- 12 beeswax baskets and 24 eggs

Morning Exercises: As before.

Lesson Content:

1 "Now Farmer Jo/e thinks about all of her neighboring families. She lives near a very small town, and everyone is very friendly. Many people have helped her in her farm work. Some neighbors even came to help her build her hen house; and if one of her hens gets outside the hen yard, a neighbor will often find it and bring it back. Farmer Jo/e really wants to share her eggs equally with all of her neighbors, but there are more neighbors than her hens lay eggs every day. Hmmm. How can Farmer Jo/e solve this problem?

[Take a little time for your child to think about this and perhaps come up with Ideas.]

Farmer Jo/e decides to give a few eggs to a few neighbors every day. If she gives eggs to a different group of neighbors every day, by the end of the week she will have shared with all of her neighbors."

2 Take out your baskets and eggs. "On Monday, Farmer Jo/e decides to give 3 eggs to 5 neighbors. How many eggs will she need to collect?" Count out one basket for each neighbor–5 baskets. Place 3 eggs in each basket. Now count the eggs you've placed in the baskets; there are 15. "Here we have given 3 eggs to each of 5 neighbors; so we've used 3 eggs 5 times. The arithmetic sentence we use to describe this is, ' Five times three is fifteen.'"

3 Continue to create other scenarios, with varying numbers of neighbors and varying numbers of eggs. Here are some examples:

Neighbors:	Eggs:	Total:
8	2	16
4	6	24
6	3	18
12	2	24
7	3	21
6	4	24

4 Always follow the process outlined in #2 above. Remember to review the process every time by saying, "So ___ times ___ equals ___. She needs ___ eggs in order to give ___ eggs to each of ___ neighbors."

5 Open your books to **page 42.** Draw a picture of Farmer Jo/e surrounded by 6 baskets, with 3 eggs in each basket. Write "18 eggs" on the bottom of the page.

6 Close your books, put away your materials, and stand for the closing verse

FIRST GRADE ARITHMETIC—TEACHER'S GUIDE

Focus: Multiplication.

Materials:
- Play ball
- Jumprope
- Copper rod
- ML Book #2
- Stick and block crayons

Morning Exercises: As before.

Lesson Content:

1. Review the previous lesson. "Where did Farmer Jo/e live? What did she want to do with her eggs? What problem did she have? How did she solve the problem?"

2. Open your books to **page 43.** Divide the page into 8 sections as before.

3. Again, we start with the upper left hand section. "Farmer Jo/e wants to give 2 eggs to each of 8 neighbors. How many eggs will she need?" Across the top of the section, draw 8 baskets, one for each neighbor. Draw two eggs in each basket. [You can draw the eggs in yellow and circle them in brown to represent the baskets.] "How many eggs does she need?" Instruct your child to count all of the eggs.

4. "This is the way we write this story in an arithmetic sentence: 8 x 2 = 16." Remind your child that we read the symbol [x] as "times," and that when you say "times," you are working with *multiplication*. We say "multiply" and "multiplication" because we want to know how many of something we will need.

5. Continue to work in each section, telling a similar story for each. Use varying numbers of eggs and neighbors as in the previous lesson.

6. Close your books, put away your materials, and stand for the closing verse

SKILLS CLASS BLOCK ONE

Modeling With Plasticine and Beeswax

WALDORF SCHOOL SUBJECTS are taught in blocks of 3-4 weeks. In First Grade, these blocks alternate between Language Arts and Arithmetic. Both of these lesson topics are recognized as such important basic skills that they are also taught in "skills classes." The work in Skills Classes generally focuses on reviewing material that has been taught in the Morning, or Main, Lesson periods. New concepts are taught in the Morning Lesson classes. *[A third topic, Nature Studies, is also part of the First Grade Curriculum. As the foundation of the Science Curriculum, this is very important material. Many Waldorf teachers thus schedule one or two Nature Study blocks as part of their Main Lesson plans. Other teachers schedule these blocks as part of the Skills Class plans.]*

Skills classes are designed to last about 40 minutes and are often scheduled later in the day than Main Lessons. Therefore, although Skills classes often include movement exercises, they do not begin with a series of movement, singing, and reciting activities as Main Lessons do.

Our Skills Classes focus on reviewing the material that has been introduced in the Main Lesson periods. The first Main Lesson block of the First Grade is Form Drawing, and the first Skills block also focuses on *form*. In this block, we work with plasticine and beeswax as well as painting and building with popsicle sticks. With these activities, the children are also learning the importance of sequence and paying attention to detail. These skills are foundational to mathematical thinking.

It is again helpful to separate a school class from customary time at home by saying a verse to begin and end the class. Here are two possibilities:

Verse to begin Skills Class [unknown author]:
Here we are with joyful hearts
Working well and working hard;
Helping gladly, quick and bold,
Bringing joy to young and old.

Verse to end Skills Class [unknown author]:
We are straight, we are strong,
We are valiant and bold-
For the Sun fills our hearts
With its life-giving gold.

Focus: "The Little Red Hen"

Materials:
- Red, yellow, brown, and orange plasticine.

[Instead of brown or yellow for your dog or cat, you might choose black. The colors you choose will partly depend on the colors in your set of plasticine.]

Before teaching this lesson, please look at illustrations of the animals you'll be modeling: chicken, pig, dog, cat. Each animal has its own particular gesture, and it's nice to capture that gesture in the plasticine.

Lesson Content:

1 Stand to say the verse to begin the Skills Class.

2 Your child sits as you stand and tell the story of "The Little Red Hen" [below].

3 "Now we're going to model the characters in this story and use our models to tell the story. We'll model two characters today. And of course, we'll start with the Little Red Hen!"

4 Animals are most easily modeled by first forming an egg shape out of your modeling material. Break off a chunk of red plasticine that easily fits in your hand. Give the child a similar chunk of plasticine.

5 "Follow me. We're going to start by making an egg from our plasticine." Plasticine is a cold material, so you will need to work with it for a little while until it becomes easily malleable. Form an egg by squeezing and rolling it in your hands, not by rolling it on a hard surface. [This is a very good activity for strengthening the hands and the connection between the brain and the hands.] Then squeeze it so that one end is smaller than the other.

6 Now you have an egg. The small end will be the head of your Hen, and the large end will be the body.

7 Push the "head" up a bit so that it is parallel with the "body." With your finger, make a depression between the head and the body and squeeze out a 'beak' from the front of the head.

8 Pull out the other end of the body until you have a nice tail shape.

9 Squeeze out a bar of clay from the bottom of the body to be the legs and feet. You'll make the legs as if they are fused together, and then squeeze the feet from the bottom of the legs.

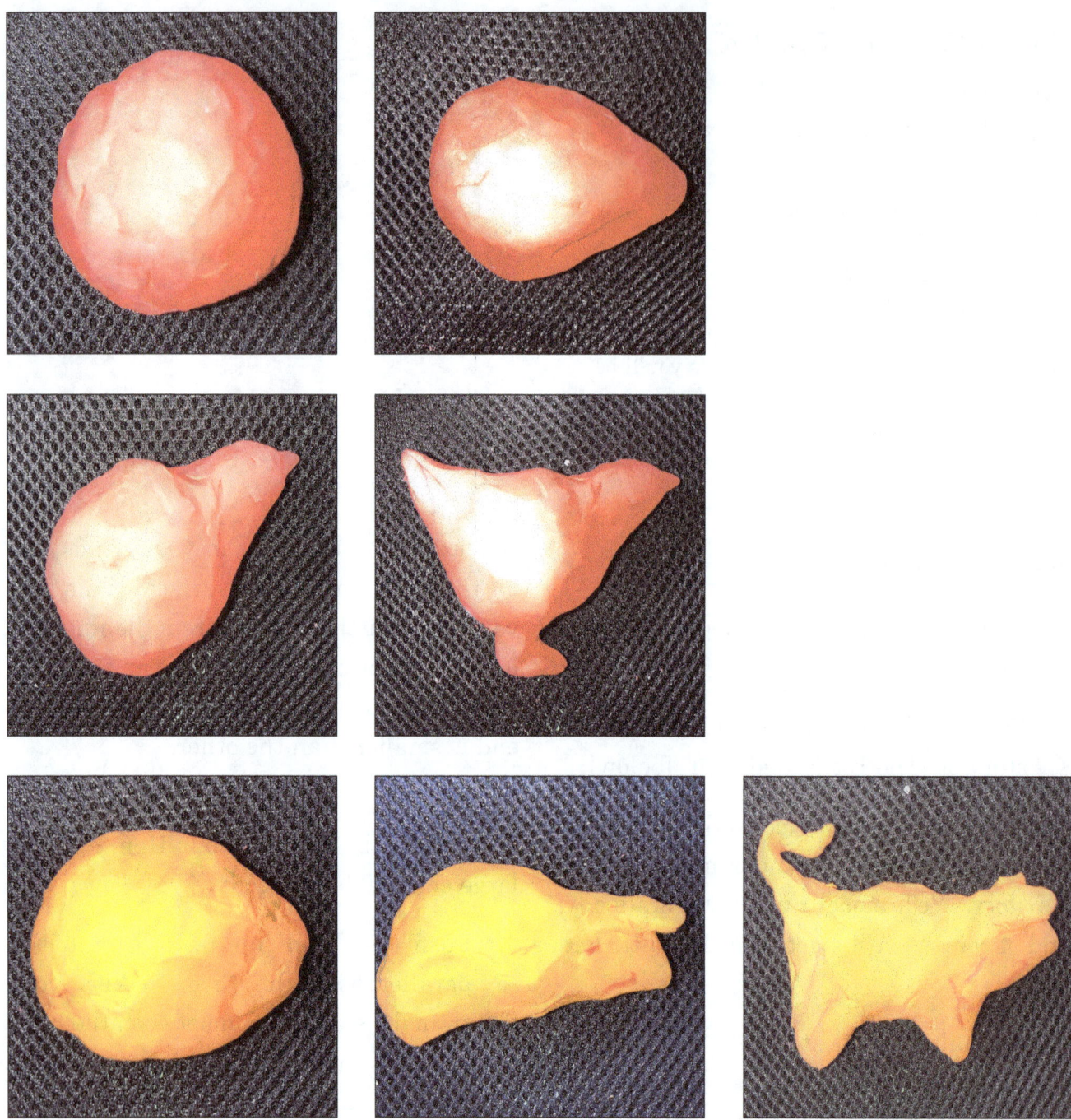

10 Now you have a Little Red Hen.

11 Break off a chunk of the orange plasticine that fits in your hand. Give your child a similar chunk. Knead the plasticine until it is warm and malleable.

12 Pigs also 'live' in an egg shape! Repeat the formation of the egg, as above. Again, the small end of the egg will become the head of the pig.

13 Unlike chickens, which carry their heads upright, pigs are always looking at the ground. Form the small end of your egg into the head and snout of the pig. Squeeze out some plasticine to form the ears, which are very characteristic of pigs.

14 Now make a barrel shape from the rest of your plasticine. Squeeze out little chunks for the four legs, which are rather like columns.

15 Squeeze out a little curlicue tail.

16 Place your animals where they will not be disturbed. Say the poem to end the lesson.

The Little Red Hen

Skills Class One-1

Traditional Children's Story

ONCE UPON A TIME there was a little Red Hen, who lived with her friends: a cat, a dog, and a pig. The little Red Hen liked to stay busy, but I'm afraid her friends were very lazy. The cat liked to sleep in the sunny window all day; the dog liked to sleep on the doormat in the shade; and the pig liked nothing else but to wallow in the cool mud by the side of the house.

One day as the little Red Hen was sweeping the path to her door, she found some grains of wheat. "Oh!" she exclaimed. "I can plant these grains of wheat, and the wheat will grow, and we can harvest it, and then we'll have flour for our bread! Who will help me plant the wheat?"

"Not I," mumbled the dog.

"Not I," yawned the cat.

"Not I," squealed the pig.

"Then I'll just plant it myself!" said the Little Red Hen. And she did.

The Little Red Hen watched as the seeds took root and grew. When weeds appeared, she asked, "Who will help me pull the weeds?"

"Not I," mumbled the dog.

"Not I," yawned the cat.

"Not I," squealed the pig.

"Then I'll pull the weeds myself!" declared the Little Red Hen. And she did.

When the wheat had grown tall and was ready to be harvested, the Little Red Hen asked, " Who will help me harvest the wheat?"

"Not I," mumbled the dog.

"Not I," yawned the cat.

"Not I," squealed the pig.

"Then I'll cut the wheat myself!" said the Little Red Hen. And she did.

The Little Red Hen cut the wheat and placed the harvested grain in her wheelbarrow. "Who will help me push the wheelbarrow to the miller's?" she asked.

"Not I," mumbled the dog.

"Not I," yawned the cat.

"Not I," squealed the pig.

"Then I'll have to do it myself!" sighed the Little Red Hen. And she did.

She pushed the wheelbarrow to the miller's and watched as he ground the wheat into flour. He placed the bags of flour into her wheelbarrow, and she pushed it home. "Now who will help me bake the bread?" she asked.

"Not I," mumbled the dog.

"Not I," yawned the cat.

"Not I," squealed the pig.

"Then I'll just do it myself!" declared the Little Red Hen.

She put on her apron and measured the flour into a bowl with the other ingredients for making bread. Then she waited until the dough was ready to bake. She placed the pan of dough into her oven. While she waited for the bread to bake, she put a clean cloth on her table and gathered flowers for the vase. When she noticed the delicious scent of the bread, she took off her apron and put a clean plate at her place at the table.

The bread smelled better and better. There is nothing like the smell of fresh, warm bread! The other animals began to gather at the door.

"Oh!" sang the Little Red Hen. "Now, who will help me eat the bread?"

"I will!" barked the dog.

"I will!" meowed the cat.

"I will!" oinked the pig.

"No!" declared the Little Red Hen. "You would not help me plant the wheat, or pull the weeds, or harvest the wheat, or push the wheelbarrow to the miller, or bake the bread. I did all that with no help. And now I will eat the bread….with no help!"

THE END

Focus: Review of "The Little Red Hen;" modeling the dog and the cat.

Materials:
- Plasticine colors for the dog and cat [brown, white,].

Lesson Content:

1. Say the verse to begin the Class.

2. Review the story of "The Little Red Hen."

3. Choose a color for modeling the dog. Break off a chunk of plasticine that will fit into your hand. Knead it until it is soft and malleable. Decide what sort of dog you will model. Talk about the features of this breed of dog: How long is its nose? What sort of ears does it have? Is its body long and skinny or short and round? How long are its legs? What sort of tail does it have?

4. Form an "egg" shape with the plasticine. The small end will be the head of the dog, and the large end will be the body.

5. Pull out the small end to form the snout of the dog. Pinch the ears out of the plasticine.

6. Shape the body for the dog you've chosen. Pinch out and form the four legs and the tail. [These instructions are for a standing dog.]

7. Now break off a chunk of the color for the cat. Again, knead the plasticine until it is soft and malleable. Form the "egg" shape. [These instructions are for a sitting cat.]

8. Pull out the small end of the "egg" to form the head and nose of the cat. Pinch the ears on the sides of the head.

9. The body of the cat will be larger at the end opposite the head, as it is sitting. The front legs do not need much detail–just dig a groove on each side of the body with your pencil and pull out the paws. We don't need to see the back legs at all.

10. Pull out the tail and wrap it around the front legs.

11. Put away your materials and say the poem to end the lesson.

136 FIRST GRADE ARITHMETIC—TEACHER'S GUIDE

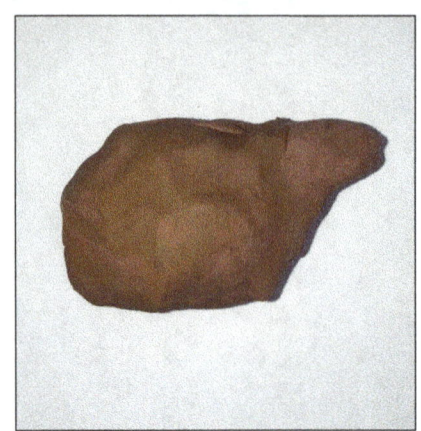

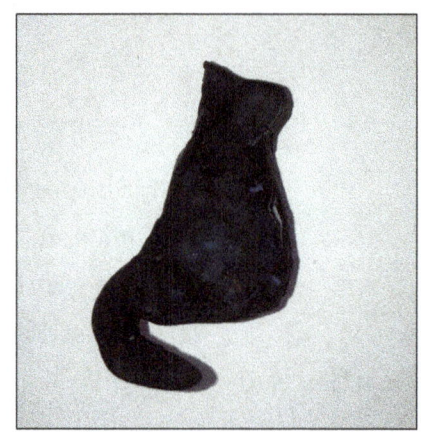

Focus: Create a background for *The Little Red Hen*.

Materials: Painting supplies:
- Painting boards
- Yellow, gold, and cobalt or ultramarine blue paints
- 1" flat paintbrush
- Watercolor paper, at least 11" x 14", and at least 130 lb. wt.
- Quart jar of clean water
- Two sponges
- Spray bottle of clean water

Lesson Content:

1 Say the verse to begin the lesson.

2 "Today we're going to paint a background for our story. Where was the Little Red Hen when she found the grains of wheat?" She was outside, sweeping the dirt in her yard to make it nice and neat. What kind of day was it? Was it raining? Was it really cold outside? The children can figure this out logically: if she was sweeping dirt outside, it must have been a dry day, and it was most likely a warm day.

3 *You will demonstrate the painting for the child, who will then paint their own painting. It is really important that you demonstrate the following steps carefully and quietly, so that your child can complete their first painting lesson successfully.*

4 Spray one side of a piece of watercolor paper; then place the paper, wet side down, on the painting board.

5 Now spray the top side of the paper. Use your sponge to "fasten" the paper in place and mop up the excess water. Start in the center of the paper and sweep your sponge to the edge of the board. Continue this motion until you have mopped up all excess water.

6 "On the day the Little Red Hen found the grains of wheat, the sun was shining brightly." Dip your brush in the water; wipe off excess water on the lip of the jar; and dip your brush into the gold paint. Wipe off any excess paint on the lip of the jar.

7 Paint a golden sun toward the center top of your paper. Dip your brush in the water and swirl it around to clean off the gold paint. Wipe off excess water on the lip of the jar.

8 Dip your brush in the yellow paint. Wipe off any excess paint on the lip of the jar. Paint yellow rays reaching out from your golden sun.

9. Dip your brush in the water and swirl it around to remove the yellow paint. Wipe your brush on the lip of the jar to remove excess water. Dip your brush in the blue paint. Wipe excess paint on the lip of the jar.

10. Now carefully paint your sky around the golden sun. Paint in soft swirls, being careful not to touch the gold or yellow paint.

11. When you have painted the entire page, around the sun, with blue, dip your brush in the water and swirl it around to remove the paint. Make sure all paint is removed from your brush. With the forefinger and middle fingers of your hand, squeeze the water from your brush and lay it aside.

12. Do not remove the paper from the paint board. Rather, pick up the board [with the painting still on it] and move it to a place where it will not be disturbed until it dries. Your brush would be happiest if you could hang it, brush side down, so that any remaining water would drip down rather than pooling on the bristles. If you cannot hang your brush, leave it lying down flat until it dries.

13. Now lead your child through the steps outlined above.

14. When your child finishes painting, show them where to place their painting board and review how to clean the brush.

15. Say the verse to end the lesson.

Focus: Creating a house for the animals.

Materials:
- Popsicle sticks
- A bottle of liquid glue, like Elmer's
- A base for your house, e.g., a piece of cardboard or a flat box

Lesson Content:

1. Say the verse to begin the lesson.

2. Start by placing two popsicle sticks in parallel lines on the cardboard. Place dots of glue on either end. Now place two more popsicle sticks, parallel to one another, on top of the glue. You have made a square.

3. Now you have some choices. You can continue this process until your house is about 5" high. Place a piece of cardboard on top for the roof. You will have a four-sided house, but it will have no door. If you'd like a door, you can use tempera or acrylic paint to paint one.

4. Alternatively, you can only place two layers of four popsicle sticks on your base. Then break a popsicle stick into four pieces. Continue to add layers to your structure; but at two of the corners, you will need to use these pieces as supports between every two popsicle sticks. This will give you a three-sided house.

5. When the glue has dried, you can use tempera or acrylic paint to paint the house, or you can leave it unpainted. It will dry overnight.

6. You can add features to your house. For example, you can use plasticine to create shrubbery, flowers, a chimney, a rug for the dog, mud for the pig, a stool for the cat, etc. [Of course, if you're really ambitious and crafty, you can figure out how to leave a window in one wall for the cat!]

7. Say the verse to end the lesson.

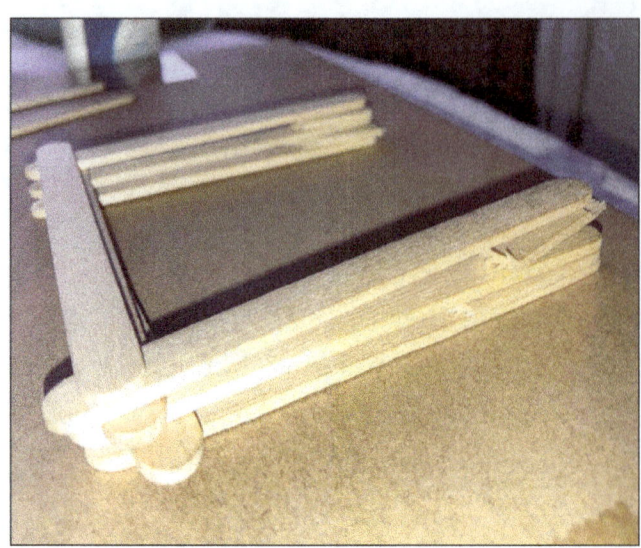

First Grade Arithmetic—Teacher's Guide

Focus: Reviewing and retelling *The Little Red Hen.*

Materials:
- Your four animals, painting, and house.

Lesson Content:

1. Say the verse to begin the lesson.

2. Prop up your background painting and position the house in front of it.

3. Review each of the animals and talk about their characters.

4. Now review the story by asking your child questions such as: Where did all the animals live? Who took care of them? What did the Little Red Hen find in the yard? What did she plan to do with the seeds? Etc.

5. Ask your child to tell you the story, using the animals and moving them around appropriately.

6. If your child does not remember one or more parts of the story, ask them a "leading question" to help them remember. The sequence of events in the story can be as long or as short as you make it. In the original story, the sequence was:

- Find the seeds/grains.
- Plant the seeds/grains.
- Water and weed them.
- Harvest them.
- Take them in the wheelbarrow to the miller, who grinds them into flour.
- Bring the flour home in the wheelbarrow.
- Make the dough.
- Bake the bread.
- Eat the bread.

At every step of the way, the Little Red Hen asks the other animals, "Who will help me…?" And the animals reply, "Not I!" every time.

7. Put away your materials and say the verse to end Skills Class.

SKILLS 1-6

Focus: Review *"The Monkey's Heart"* and model a monkey from beeswax.

Materials:
- Half a stick of brown or black beeswax

Lesson Content:

1 Say the verse to begin the lesson.

2 Take the beeswax and hold it in your hands in order to warm it. Review *The Monkey's Heart*. One way to review a story is to start the story as you originally told it. After two or three sentences, stop and encourage your child to continue the story. The goal is for the child to tell as much of the story as possible with, at most, occasional prompts from the adult.

3 This story has several episodes:

a. The monkey eats fruit from a large tree.
b. The monkey drops fruit into the sea for the shark to eat.
c. The shark invites the monkey to come to its home.
d. Monkey at first refuses, but then agrees.
e. Shark swims through the sea with Monkey on its back.
f. Shark admits that the Chief Shark needs a monkey's heart in order to heal.
g. Monkey regrets that it left its heart at home, so they turn around and swim back to the tree.
h. Monkey climbs the tree while Shark waits.
i. Monkey laughs at Shark for believing it would return with Shark to the sea.

4 Again, this story contains moral elements that are worthy of discussion. Was Shark telling the truth? Was Monkey telling the truth? Was either of them justified in being untruthful?

5 By now, your beeswax should be warm enough to model. Talk about monkeys– what is their customary posture? How do they move? How would a monkey sit in a tree? Invite your child to pretend to be a monkey, moving and sitting as monkeys often do.

6 Gently press the beeswax into an oval, as you did with the plasticine. Pull the small end out a bit to form the rounded head of the monkey. [NOTE: Your beeswax should be very warm and soft.]

7 Pinch and pull out beeswax from each end of the remaining, larger end of the oval, to form the limbs of the monkey.

8 Pinch and pull out beeswax to form the long tail of the monkey. Model your monkey so it is sitting: back limbs bent at the knees, and forelimbs bent as if they are holding a piece of fruit. [You can even model a fruit out of a little red beeswax.]

9 Place your monkey in a safe place, and say the verse to end the Skills Class.

SKILLS 1-7

Focus: Modeling the Shark from beeswax.

Materials:
- One stick of white beeswax, and a little black or blue beeswax.

Lesson Content:

1. Say the verse to begin the lesson.

2. Hold the beeswax in your hands to warm it. While warming the beeswax, talk about the role of the Shark in the story. What do you know about the personality of sharks in general? What is special about sharks? What do they eat? Where do they live? How big are sharks? How many different kinds of sharks can you think of? [This might lead to a certain amount of research on the part of the adult. Visit the children's section of your local library and look for books, with nice pictures for your child, on sharks and monkeys.]

3. When the beeswax is warm, work with it for several minutes in order to blend the black into the white, thus forming a nice gray or pale blue. [NOTE: Your beeswax should be very warm and soft. Feel free to just choose one color for your shark rather than trying to blend the colors.]

4. Again, form an oval with the beeswax. This time, however, pull the beeswax out from the center into a long shark form. Remember the fins and the tail. Be sure to leave a space on the shark's back so the monkey will have a place to sit.

5. Seat your monkey on top of the shark, and put them in a safe place.

6. Say the verse to end the Skills Class.

Focus: Creating the environment for your story.

Materials:
- Your monkey and shark, and other supplies as noted.

Lesson Content:

1 Say the verse to begin the lesson.

2 In this lesson, we will be making or finding a tree for the monkey and a sea for the shark. If you are able to go outside, do! Search for a sturdy and interesting branch that will serve as the tree for the monkey. You might also look for rocks, moss, and smaller clusters of leaves to place around your "tree."

3 Bring these things back inside. Use a clump of plasticine to anchor your "tree." Place the natural items you've gathered onto a flat piece of cardboard.

4 Now it's time to make the sea. You have several choices here:

  a. If you have blue fabric, your child can simply cut a piece that will fit your cardboard base. [Felt is really a nice choice for fabric.]

  b. Your child can paint a blue sea. If you decide on this option, follow the procedure in the previous lesson.

  c. Your child can cut a piece of blue construction paper or cardstock into a

shape that will fit your base. If you decide on this option, it is nice to use two or more shades of blue paper. Choose one shade for the bottom piece, then cut smaller pieces out of the other shade and glue them on top. This will be much more interesting if you round all the corners of the base and cut the smaller pieces into various "swirly" shapes. These will represent the warmer water that swirls within the cold ocean.

5. Now you can set up your "diorama" and enact the story.

6. Put your materials in a safe place, and say the verse to end the Skills Class.

SKILLS CLASS BLOCK TWO

Solidifying the Understanding of Numbers

THE FIRST SKILLS BLOCK focused on modeling with plasticine and beeswax. The second skills block will focus on solidifying the child's understanding of numbers 1-12 as well as rote counting from one to one hundred. We'll paint numbers 1-12, practice subitizing, and jump rope. In addition, we'll make a collage for each number. This will give the child an opportunity to practice working with scissors and glue.

Painting the numbers gives the child an opportunity to practice the forms of the numbers using the larger muscles of the hand and arm. These six paintings [two numbers each] will be easy to hang on the walls so the child will be reminded daily of their forms, their order, and their meaning.

Subitizing is a foundational skill in the growth of understanding and working with arithmetical operations. Most adults can see a set of, for example, four apples, and know that it is four without having to count the apples one by one. When a child first learns numbers, however, this ability is not necessarily automatic. [In fact, one recent understanding is that children who struggle in arithmetic do so because this skill, among others, is underdeveloped.] Imagine what your life would be like if you had to start at "one" and count up every time you met a set of objects!

When we learn basic patterns for items, we learn to subitize faster. For example, we might see three dots as the points of a triangle, or four dots as the points of a square. Then, if we see three dots and four dots pictured in those patterns, we'll know immediately that there are seven dots.

In the appendix, we have provided the addresses of online sites for dot cards as well as instructions for making your own. Instructions for using them with your child are provided in the first lesson of the second skills block.

Jump rope! If you've forgotten how, this is the time to remember it! In Waldorf schools, we start with a rope that is swung by adults so the children can jump over it. Commercially available jump ropes are often too flimsy to be of real use. You can buy lengths of heavy, ¾" rope by the foot or yard at a hardware store. While you're there, buy a long rope [10' - 12'] for your beginning jumper as well as a shorter rope for when your child is ready to jump independently. [Measure how long the shorter rope will need to be by having your child loosely hold the two ends of a string at waist height while stepping on the middle. This should be long enough to comfortably go over your child's head while jumping.]

The steps for learning how to jump rope are:

1. Lay the long rope flat on the floor and have your child practice jumping over it until they are comfortable running at the rope and jumping over it [rather than stopping at the rope and jumping only after hesitating].
2. As your child becomes more comfortable with jumping over the rope, gradually raise the level of the rope, inch by inch, until it is about 4" high.
3. When your child is comfortable with the raised rope, gently swing it back and forth. This will mean tying one end of the long rope to a fence post, etc., as you will likely be the only adult available to hold the rope.
4. When your child is comfortable jumping over the swinging rope, have the child jump over and back. Children love to count how many times they can jump back and forth.
5. When your child is very comfortable with this, swing the rope over the child's head and then under the child's feet. At first, you might be able to do this only once, but gradually build up the number of times your child will allow you to swing the rope all the way around.
6. Give your child the smaller rope and show them how to swing it themselves.

Focus: Review of numbers 1-12, counting to 100, preparation for painting numbers.

Materials:
- Dot cards, 1 - 5 dots
- Jump rope
- Painting supplies:
  Watercolor paper, 2 pieces
  Yellow paint
  1" flat brush
  Quart jar of water
  Sponge
  *Optional: magazines, catalogs, glue stick, scissors, heavy printer paper [letter size]

Lesson Content:

1. Stand to say the verse to begin the lesson.

2. Engage your child in the appropriate level of jump roping skills. If your child is just beginning, use the long, thick rope and follow the sequence outlined in the Introduction to Skills Block Two. Have your child jump over the rope at least 12 times while counting 1 - 12. Gradually add the number of jumps with counting: first, extend the activity to 20, then to 25, and so on. Practice this every day! The goal is to be able to count to 100 while jumping independently with a child-sized jump rope.

3. When your child is seated again, hold up your fingers in combinations from 1 - 5, asking your child to tell you how many fingers they see.

4. Now say a number, 1-10, and ask your child to hold up that many fingers.

5. Play the "dot card game." Hold up one card at a time and ask your child to tell you how many dots they see. Be sure to hold the cards up in a random order. When your child understands the game, increase your speed. You are teaching your child to "subitize," which means to recognize a quantity quickly without having to count the parts.

6. Next, you will paint a piece of watercolor paper with yellow paint. This painting will be the background for the next lesson, when you will paint numbers. First, set up your painting station. While your child watches, paint one piece of watercolor paper with yellow paint, gently moving your brush from left to right in a straight line and continuing until the paper is completely full.

7. Now it's your child's turn to paint. They should follow your process.

8 Place your yellow, wet papers where they will not be disturbed. Put away your materials.

9 Stand to say the verse to end the Skills Class.

Optional: Take out the magazines, catalogs, glue, scissors, and printer paper. Explain to your child that you will be creating a "Book of Numbers." Each number will have a page, so start with one piece of printer paper. [You will, of course, have previewed the magazines and catalogs before the lesson, to ensure that the pictures are child-friendly.]

Ask your child to hold up ONE finger and to find a picture of ONE thing to glue on the page.

Follow that procedure with TWO and THREE.

Place the three pages in a safe place so the glue will dry. You might place a heavy book on top of each so the pages won't buckle.

This activity is especially helpful if your child has not had much experience with scissors and cutting, or if your child finds working with scissors a challenge.

Focus: Counting, subitizing, painting numbers 1 and 2.

Materials:
- Jump rope
- Dot cards
- Watercolor paper: 2 yellow papers from yesterday, plus 2 clean sheets
- Yellow, blue, and red paints
- 1" flat brush
- Quart jar of water
- Sponge paint boards
- ML Book #1
- Block and stick crayons

[*Optional: magazines, etc., as before.]

Lesson Content:

1. Stand to say the verse to begin the lesson.

2. Jump rope as previously.

3. Practice subitizing with fingers and dot cards as previously.

4. Open your ML book #1 to **page 19**. As you will have introduced the alphabet in your second ML block language arts classes, your child will be familiar with it. You might even have posted the alphabet on a wall. At any rate, in this "extra" set of Skills activities, you will be talking about letters and geometric figures that represent each number, 1 - 12.

5. Explain to your child that you will be drawing geometric figures, like triangles and squares. Now ask your child if there are any geometric figures that use only one straight line. Well, yes–it is a straight line!

6. On the upper area of **page 19, draw a straight vertical line and a straight horizontal line**.

7. Now ask your child if any letter of the alphabet can be written with only one straight line.

8. These letters are: B, D, G, J [with a "cap"], P, Q, U [with a line on the right].

9. On **page 19,** draw a bar of color across the bottom of the page. Within this bar, practice drawing your ONE.

10. On top of the bar, write your letters as noted in #7, above.

11. Now ask, are there any geometric figures that have TWO straight lines? The answer is no–two lines can be joined at one end, but not at the other. However, parallel lines are really important and easily seen in our lives. On **page 20,** draw two vertical parallel lines on the left side of the page, and two horizontal parallel

lines across the middle of the page. Ask you child if they can think of where they might have seen lines that look like this. Some possible responses are: highway lines; the lines of a door, window, or house; the lines at the top and bottom of a piece of paper, etc.

12 Draw a bar of color across the bottom of the page, and within this bar, practice writing your TWOs.

13 On top of this line, after asking your child to find them, draw the letters that have TWO straight lines in them. [These are: L T V Y R X. Including "Y" will depend on how you draw this letter, as it could also be seen as having 3 straight lines.]

14 Put your ML book and crayons away, and take out the piece of yellow paper you painted in the previous lesson.

15 Lay your yellow paper flat. You'll paint two numbers on this paper, so be sure to space them as shown in the photograph. Paint the number **ONE** as a straight, vertical line, with the flat side of your brush and the blue paint. Paint the number slowly and carefully as your child watches.

16 Now your child paints their **ONE.**

17 Dip your brush in the water and swirl it around until the blue paint is gone. Wipe off any excess water on the lip of the jar. Now dip your brush in the red paint and carefully paint the number **TWO** next to the one. Again, keep the flat side of the brush down.

18 By the end of this series of Skills Classes, you will have painted six paintings with two numbers on each, 1 - 12. Odd numbers [1, 3, 5, 7, 9, 11] will be in blue; and even numbers [2, 4, 6, 8, 10, 12] will be in red.

19 Dip your brush in the water and swirl it around until the red paint is gone. Wipe off excess water on the lip of the jar.

20 Now dip your brush into the yellow paint and paint your clean sheet yellow as before.

21 It's your child's turn to paint their clean sheet yellow, as you have done.

22 Put your wet paintings in a safe place and put away your materials.

23 Stand and say the verse to end the class.

Optional: Follow the procedure outlined in the previous lesson for the numbers FOUR and FIVE.

First Grade Arithmetic—Teacher's Guide

SKILLS 2-3

Focus: Counting, subitizing, painting numbers 3 and 4.

Materials:
- Jump rope
- Dot cards
- Painting suppliess
- ML Book #1
- Block and stick crayons

*[*Optional: magazines, etc., as before.]*

Lesson Content:

1. Stand and say the verse to begin the lesson.

2. Jump rope as before.

3. Practice subitizing with fingers and dot cards as before.

4. Open your ML#1 book to **page 21.**

5. Ask your child what geometric figure has THREE lines. They will most likely respond correctly with "a triangle."

6. What letters have THREE straight lines? [H A F Z K N I {if drawn with "feet" and a "cap"} and Y {depending on how you draw it}]

7. Draw a bar of color across the bottom of the page. Within the bar, practice your THREEs.

8. On top of the bar, draw the letters listed above.

9. Draw THREE triangles in the remaining space on the page.

10. Now turn to **page 22.** Ask your child what shapes have FOUR straight lines. They will likely respond correctly with "square" and "rectangle." Other quadrilaterals are also acceptable! You can draw parallelograms, rhombuses, and trapezoids. Children love big words.

11. Which letters have FOUR straight lines? [E M W]

12. Across the bottom of the page, draw a bar of color. Within the bar, practice your FOURs.

13. On top of the bar, draw the letters listed above.

14. In the remaining space, draw FOUR quadrilaterals. [If you like, you can explain that "quadrilateral" is a word that comes from the ancient language of Latin. "Quad" means "four," and "lateral" means "side."]

FIRST GRADE ARITHMETIC—TEACHER'S GUIDE   155

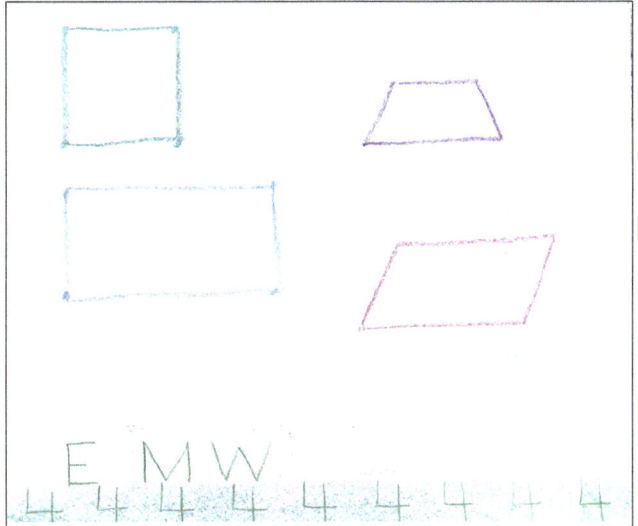

15 Put away your ML book and crayons. Take out the piece of yellow paper you painted in the previous lesson. Paint the numbers 3 and 4 on your dry yellow paper as before.

16 Paint new yellow pages as before.

17 Put your wet papers in a safe place and put away your materials.

18 Stand to say the verse to end the lesson.

Optional: Follow the procedure outlined in the first lesson for the numbers SIX and SEVEN.

Focus: Counting, subitizing, painting numbers 5 and 6.

Materials:
- Jump rope
- Dot cards
- Painting suppliess
- ML Book #1
- Block and stick crayons

[*Optional: magazines, etc., as before.]

Lesson Content:

1. Stand to say the verse to begin the lesson.

2. Practice jumping and counting as before.

3. Practice subitizing with fingers and dot cards as before.

4. Ask your child, "Are there letters which are made with FIVE straight lines?" Wait for your child to look at all of the letters you have displayed. The correct response is, "No."

5. Are there any letters we haven't used? Yes, as you survey the entire alphabet, there are three letters that have not been organized as containing straight lines: C, O, S. When your child identifies these three letters, ask them what these letters have in common. [They are composed only of curved lines.]

6. Turn to **page 23** in the ML #1 Book. Draw a bar of color across the bottom of the page, and practice your FIVEs within it.

7. Now ask your child if they know any geometric figures that have FIVE sides. A *pentagon* [as in THE Pentagon outside Washington, DC] has five sides. A *pentagram* is a five-pointed star. As your child has already worked with the star shape, it might be easiest to begin with the child making the star with their body: arms and legs outstretched, head upright. Draw the star as a pentagram, then connect the points to create the pentagon. Each has five lines.

8. Now instruct your child to draw more forms with five lines. These forms can be any shape as long as the five lines are connected in some way.

9. Put away the ML books and crayons, and take out the yellow paper you painted in the previous lesson. Paint the numbers **FIVE** and **SIX** on your dry yellow paper as before.

10. Paint new yellow pages as before.

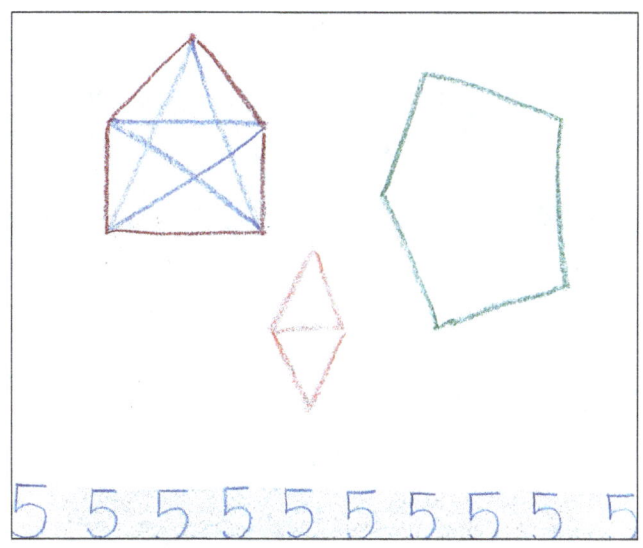

11. Place your wet papers in a safe place and put away your materials.

12. Stand to say the verse to end the lesson.

*Optional: Follow the procedure outlined in the first lesson for number EIGHT.

SKILLS 2-5

Focus: Counting, subitizing, painting numbers 7 and 8.

Materials:
- Jump rope
- Dot cards
- Painting suppliess
- ML Book #1
- Block and stick crayons

[*Optional: magazines, etc., as before.]

Lesson Content:

1. Stand to say the verse to begin the lesson.

2. Jump rope and count as before. As your child becomes comfortable with counting and jumping, increase your expectations.

3. Practice subitizing with fingers and dot cards.

4. On the dry yellow paper, paint the numbers **SEVEN** and **EIGHT.** Paint two new pieces of paper with the yellow paint, in preparation for the next class.

Place your wet papers in a safe place and put away your painting materials.

5. Open your book to **page 24.** Draw a bar of color across the bottom of the page and practice the SIXes within it.

6. Tell your child that the form with SIX sides is called a hexagon. The six-pointed star is called a *hexagram*. [You can explain that "hex" is from the ancient Greek word for "six."] The easiest way to draw a *hexagram* is to draw two triangles: one, "flat side down," and the other, on top of the first, "flat side up."

7. Draw a hexagon by following these steps:

   a. Draw two parallel horizontal lines, about 3" long and 5" apart.

   b. Out from the middle of the space between the lines, and on each side of that space, draw a dot.

   c. Draw lines connecting the dots to the ends of the lines.

   d. You can also draw a hexagon by first drawing the hexagram and then connecting the points with straight lines. It is good practice to draw the hexagon using both methods. And children usually love learning how to do this!

FIRST GRADE ARITHMETIC—TEACHER'S GUIDE 159

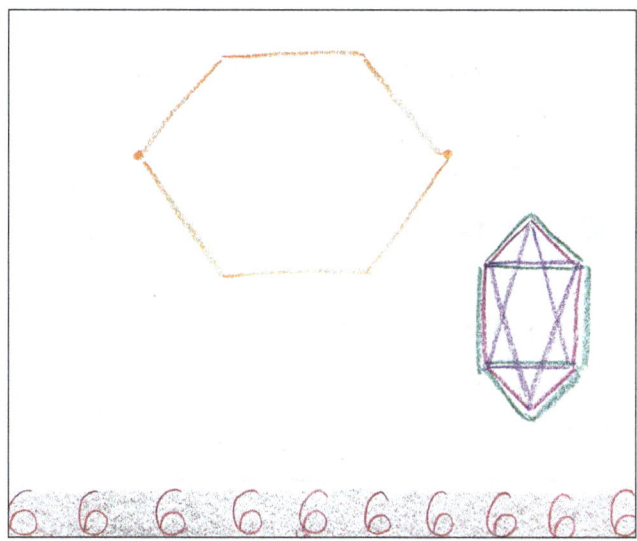

8 Now instruct your child to draw more six-sided forms, in any configuration they choose.

9 Stand to say the verse to end the lesson.

Optional: Follow the procedure outlined in the first lesson for the number NINE.

SKILLS 2-6

Focus: Counting, subitizing, painting numbers 9 and 10, collage for numbers 4-6.

Materials:
- Jump rope
- Dot cards
- Painting suppliess
- ML Book #1
- Block and stick crayons

[*Optional: magazines, etc., as in the first lesson.]

Lesson Content:

1 Stand to say the verse to begin the lesson.

2 Practice jumping rope and counting as before.

3 Practice subitizing as before. By now, you should be showing the numbers on the cards and with your fingers with much less time in between.

4 Paint the numbers **NINE** and **TEN** on your dry yellow pages.

5 Paint the yellow pages you will use in the next lesson.

6 Place your wet pages in a safe place and put away your painting supplies.

7 Open your books to **page 25**.

8 Draw a bar of color across the bottom of the page and practice the SEVENs.

9 Now you will draw a seven-sided geometric form, called a *septagon* or a *heptagon*. First draw a large circle.

10 Next, using an erasable pencil, place 7 dots, equally spaced, on the perimeter. [Your child will probably need help with this.]

11 Connect the dots with straight lines.

12 Now turn to **page 26**. Draw a bar of color across the bottom of the page and practice the EIGHTs.

13 On this page, you will draw an octagon. The easiest way to do this is to first draw a square, with each side 3-4" long.

14 With another color, draw lines creating triangles at each corner, such that you have 4 triangles with bases of 1.5-2" long. Using that same color, connect the points of the baselines. Now you have 8 lines, each 1.5-2" long.

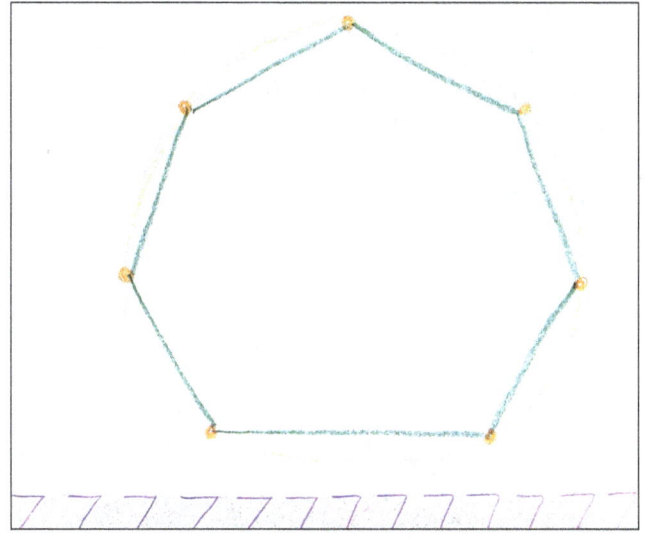

15. Another way to draw an octagon is to draw a circle 5-6" in diameter. Place a dot in the center of the top of the perimeter, and another dot on the perimeter directly below that one. Then place two dots at the diameter line on each side; and four dots, each in between two of the others. You now have a circle with EIGHT dots equally spaced on its perimeter. Connect the dots with straight lines.

16. It is really good practice to draw both ways to create an octagon, and children usually love it!

17. Take out the collage materials. You'll create collages for the numbers 4, 5, and 6 today.

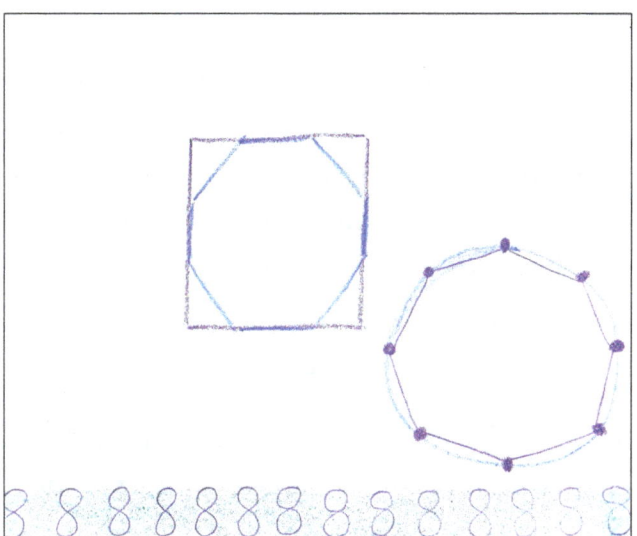

18. Place your collage pages in a safe place and put away all of your materials.

19. Stand to say the verse to end the lesson.

[*Optional: Follow the procedure outlined in the first lesson for the number TEN.]

SKILLS 2-7

Focus: Counting, subitizing, painting numbers 11 and 12, collages for numbers 7, 8, 9.

Materials:
- Jump rope
- Dot cards
- Counting stones
- Painting suppliess
- ML Book #1
- Block and stick crayons
- Colored pencils

[*Optional: magazines, etc., as in the first lesson.]

Lesson Content:

1. Jump rope and count as before.

2. Practice subitizing with dot cards and fingers as before.

3. Instruct your child to count out 12 counting stones. As you say a number, instruct your child to separate out that number of stones.

4. Paint the numbers **ELEVEN** and **TWELVE** on the dry yellow paper from the previous lesson.

5. Place your paintings in a safe place and put away your materials.

6. Open your book to **page 27**. Draw a bar of color across the bottom of the page and practice writing **NINEs**.

7. With your golden crayon, draw a large circle in the space above the nines.

8. You'll need three more stick crayons for this drawing. Your drawing will be more pleasing if you choose three friendly colors, like orange, vermillion, and carmine. With the lightest crayon, draw an equilateral [equal-sided] triangle inside the circle, starting at the center top of the circle.

9. On the perimeter of the circle [and using a graphite pencil], place two dots between every two points of the triangle. Superimpose two more equilateral triangles, using the remaining two crayon colors. With one color, connect dots #2, #5, and #8. With the other color, connect dots #3, #6, and #9.

10. You should now have a circle with nine points. Connect the points by drawing straight lines from one to the other with your red/carmine pencil.

11. Close your books and stand to say the closing verse.

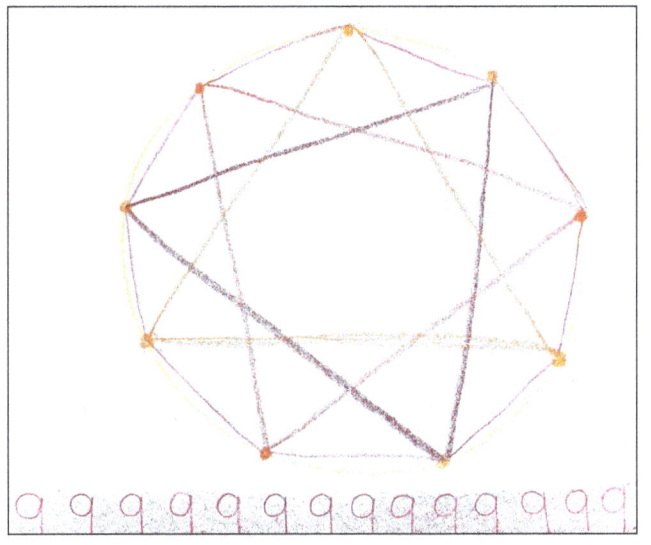

*Optional: Following the procedure from the first lesson, make a collage to represent the number ELEVEN.

Focus: Jumping rope, counting, subitizing, collages for numbers 10, 11, 12.

Materials:
- Jump rope
- Dot cards
- Counting stoness
- ML Book #1
- Block and stick crayons
- Colored pencils

[*Optional: magazines, etc., as in the first lesson.]

Lesson Content:

1. Jump rope and count as before.

2. Practice subitizing and counting as before, with dot cards, fingers, and stones.

3. Open your books to **page 28.** Explain to your child that there are forms for the numbers 10 and 11, but that they are very difficult to draw without special tools. However, you ARE going to draw a form for the number TWELVE.

4. Draw the largest circle you can on page 28. This form is created by drawing 4 overlapping triangles, and you will follow the same basic procedure that you followed in drawing the nine-form. However, even though this form has 12 sides, it is actually easier to draw than the 9-sided form.

5. First, using an erasable pencil, draw a dot at the center top of the perimeter.

6. Now draw a dot directly beneath that first dot, at the center bottom of the perimeter.

7. Draw a dot at the center left of the perimeter, and another at the center right.

8. You now have four dots, equally spaced, on the perimeter of the circle. Between each set of two dots, draw two more equally spaced dots.

9. When you finish that, you will have 12 dots equally spaced on the perimeter of the circle.

10. clockwise, connect dots #1, #5, and #9 with straight lines. That is your first triangle.

11. With a different color, connect dots #2, #6, and #10.

12. With a different color, connect dots #3, #7, and #11.

13 With yet another color, connect dots #4, #8, and #12.

14 You have created a form by overlapping four triangles. There is a large open space in the center. Shade it in with a golden crayon. Using the colors that you chose to draw each triangle, shade in the parts of each triangle that lie outside that golden center.

15 Connect all twelve dots with straight lines going around the circle. You have now created 12 small triangles. Shade these in with the golden.

16 You've also created 12 shallow domes going around the circle. Shade those in with the same color you used to draw the original circle.

17 Close your books, put away your materials, and say the closing verse.

***This form will be very lovely if you choose harmonious colors for your drawing. For example, you can draw the circle with red, and the triangles with red, vermillion, orange, and violet [as I did in the example]. Most of these colors are classified as "warm" colors, and they harmonize quite beautifully with the golden.

*Optional: Follow the procedure from the first lesson to make a collage representing the number TWELVE. Place your collage pages in a safe place and put away your supplies. After these pages dry, add them to the stack of collage pages you made previously. Mix up the order of the pages and ask your child to put them in order.

Punch three holes in the side of your pages, and place them in a 3-hole binder.

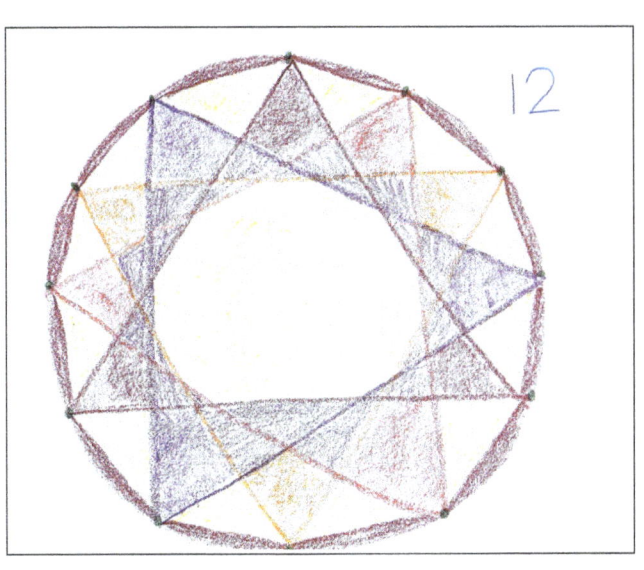

SKILLS CLASS BLOCK THREE

Beginning to See With Hands and Fingers

THIS SKILLS BLOCK BEGINS with a combination of form drawing and form modeling. The form drawing exercises are derived from a series developed by Audrey McAllen to support young children in their learning journey. Although the forms *look* very simple, drawing them requires real focus and motor control. The modeling exercises were developed by Hella Loewe. This teacher recognized that the children of today often do not have the opportunity to create things with their hands. Think about it—our world is full of buttons that, when pushed, seem to be magical. But the person who pushes the button simply watches the consequences rather than being physically engaged. Working with one's hands and seeing the concrete results is remarkably satisfying and soothing. In addition, the muscles of the child's hands are strengthened and the child can see their own creative power. One could say that the child begins to "see" with their hands and fingers.

The second set of exercises in this Skills Block return to the four operations of arithmetic. Here we discover the addition, division, subtraction, and multiplication factors

SKILLS 3-1

Focus: Subitizing, counting by twos, form drawing.

Materials:
- Dot cards, with dots arranged in specific arrangements
- Counting stoness
- Jump rope
- ML Book #1
- Block and stick crayons

Lesson Content:

1. Say the verse to begin the lesson.

2. By this time, your child is probably very comfortable jumping rope. Instruct them to jump while counting to 100, whispering on odd numbers and shouting on even numbers. If your child is already comfortable counting all the numbers, 1 - 100, instruct them to leave out the odd numbers and count only the even numbers.

3. Your child returns to their desk and sits. Show your dot cards and talk about the different arrangements of the dots. Give your child a blank card and ask them to create various arrangements with their counting stones on the card.

   a. ONE dot can be anywhere on the card.
   b. TWO dots can be in corners or in the center of the card, horizontal or vertical.
   c. THREE dots can be in corners; in the center of the card, horizontal or vertical; or arranged as a triangle.
   d. FOUR dots can be in corners; in the center of the card, horizontal or vertical; or arranged as a square in the center of the card.

4. Now shuffle the cards and show them one at a time, asking your child to name the number of dots.

5. Shuffle them again. Hold up two cards simultaneously and ask your child to name the total number of dots. By doing this, you are helping your child learn a more complex advantage of subitizing–combining the shapes made by dots/items without consciously counting the dots/items.

6. Put those materials away. Now ask your child to stand up as straight and tall as they can, stepping behind their chair and pushing it under the desk. Ask your child to walk across the room and back, in very straight lines.

FIRST GRADE ARITHMETIC—TEACHER'S GUIDE 169

9 Now demonstrate actually drawing the blue bars across the page. You'll draw six bars of blue, being careful to draw them in straight lines with no white showing in between.

10 With your red block crayon on the medium side, draw bars of color across the rest of the page.

11 Close your books and put away your materials.

12 Say the verse to end the lesson.

7 Now stand next to your child and, with your dominant hand [and the child using their dominant hand], "draw" a straight line in the air. Repeat this motion several times.

8 Open your ML Book #1 to **page 29.** You will need a blue block crayon and a red block crayon. Demonstrate "drawing" straight lines in the air, across the page with the blue crayon on the medium side.

SKILLS 3-2

Focus: Subitizing, counting by twos, form drawing.

Materials:
- Jump rope
- Dot cards
- Counting stoness
- ML Book #1
- Block and stick crayons

Lesson Content:

1. Say the verse to begin the lesson.

2. Jump rope as in the previous lesson. If your child is comfortable counting by twos without speaking the odd numbers, ask them to count by fives. Children usually love to count by fives, and they learn how to do this by rote. Count and jump along with your child, as this will help them learn the counting pattern.

3. Give your child the blank dot card. As you say a number, 1 - 4, instruct your child to create patterns for that number with the counting stones on their card.

4. Put away the counting stones and blank card, and hold up combinations of your printed dot cards. Change the combinations as quickly as your child can comfortably name the combinations. You want to challenge your child by going quickly, but not so quickly that your child becomes frustrated.

5. Put away these materials and take out your ML Book #1, as well as your red and blue block crayons. Open the book to **page 30**.

6. This form is similar to that in the previous lesson. "Draw" straight lines in the air above your page with the blue crayon. Then demonstrate for your child drawing a bar of blue across your page, followed by a bar of red.

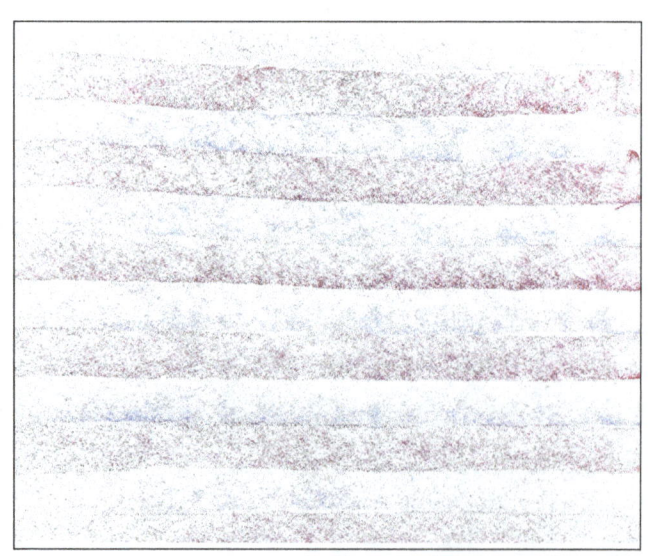

7 Instruct your child to "draw" straight lines in the air three times above their page, and then to follow your example. Continue to draw alternating bars of color until the page is full. Keep your bars as straight as you can, and try not to let any white show through.

8 Put away your materials and say the verse to end the lesson.

SKILLS 3-3

Focus: Counting by 2s, 5s, and 10s; form drawing; modeling the sphere and the ovoid forms.

Materials:
- Jump rope
- Plasticine, two sticks of any color
- ML Book #1
- Block and stick crayons

Lesson Content:

1. Stand to say the opening verse.

2. Instruct your child to jump rope, counting to 36 by 2s and to 100 by 5s and 10s. If your child is not comfortable counting independently, count with them.

3. Return to the desk and sit. Give your child one stick of plasticine and keep the other. In our last Skills block, we drew many geometric forms, and today, we're going to use our plasticine to model some forms. "First, think about the Sun. What form is the Sun? Yes, it's round. When something is round and flat, we call it a *circle*. But when something is round and fits into your hands, we call it a *sphere*. Let's make spheres with our plasticine."

4. Model a sphere with your hands while your child observes. Do not roll it on the desk, but use only your hands in modeling. When your sphere is complete, ask your child to create their sphere.

5. Now flatten your sphere. Using only your fingertips, model the sphere again. At the start of this activity, your child is observing; but they may join with their plasticine spheres at any time.

6. "Now our sphere is stretching out into the world. Perhaps it wants to go exploring!" As your child observes, pull one end of the sphere out into what may look like a nose, while keeping the other end round. "We've made a new shape. This one is called an *ovoid*."

7. Squish the plasticine and put it away. Take out your ML book and turn to **page 31**. "Now we're going to continue with the form we drew in our last lesson, but this time it will be a little different."

8. As your child observes, with the large side your blue block crayon, draw a straight bar of color across the top of the page. Using the medium side of the red block crayon as a measure, skip that space

FIRST GRADE ARITHMETIC—TEACHER'S GUIDE 173

and draw another blue bar of color, this time with the medium side of the crayon. Continue to do this until you have drawn 5 bars of blue. [The bottom bar will probably need to be drawn with the large side of the block crayon in order to fill the space.] You will also have 4 spaces.

9 Now instruct your child to draw blue bars on their page as you did on yours. When they have finished, instruct them to watch as you complete the next part of the form. With your red stick crayon, draw a wavy line across the page in the first clear space.

10 Continue to draw these wavy bars of color until all four clear spaces contain them. Now instruct your child to draw their own red waves.

11 Put away your materials and stand for the closing verse.

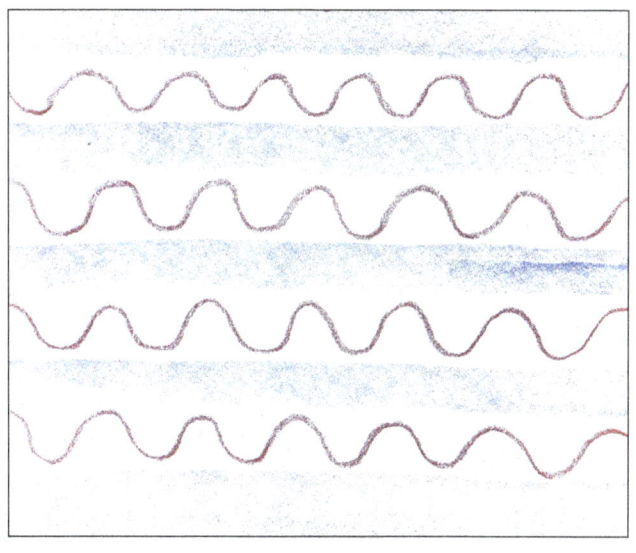

SKILLS 3-4

Focus: Counting; form drawing; modeling.

Materials:
- Jump rope
- Plasticine
- ML Book #1
- Block and stick crayons

Lesson Content:

1. Stand to say the opening verse.

2. Continue to practice counting by 2s, 5s, and 10s with your jump rope.

3. Take out the plasticine. Using your fingertips, model the sphere. When your child has completed the sphere, ask, "What happens next?" If your child does not remember, remind them that next, we model the ovoid. Ask your child to do this—it will give you a good idea of your child's memory of the form. If your child does not know how to create the ovoid, demonstrate with your plasticine. But do not make the ovoid for the child, as it is important for the child to create their own.

4. Now ask your child to put down their ovoid and hold up their hand. Using your own hand as a model, show the child the rounded area at the base of the thumb. "This part of your hand is called 'the ball' of your hand. Feel it on both of your hands. Can you feel a roundness?" [This might be easier to perceive when the hand is relaxed. The ball of the hand makes a curved form toward the center of the palm.]

5. "The ball of the hand is a tool that we're going to use in creating our next form." Demonstrate holding the ovoid in your non-dominant hand. Now, with the ball of the other hand, press into the center of the ovoid, making a hollow. Move the entire thumb joint back and forth in the hollow, thus creating a beautifully smooth transitional area between the two mounds of the ovoid. "Look! We've taken a 'one' and made a 'two!'"

6. Instruct your child to make their own 'two.' Experiment with your form, modeling the 'mounds' to create both symmetrical and asymmetrical forms [but always as 'two'].

First Grade Arithmetic—Teacher's Guide 175

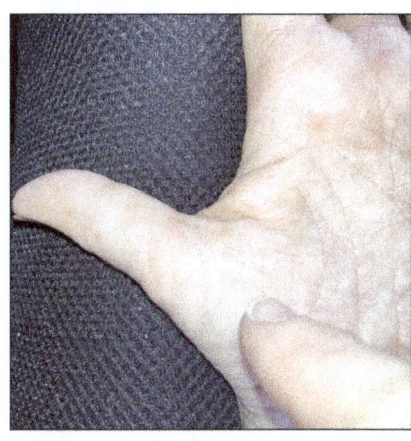 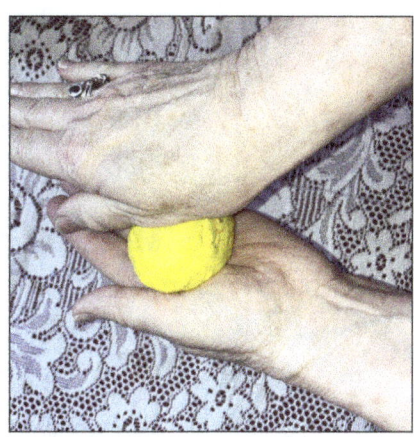 

7 Now put away your plasticine and take out your ML books. Turn to **page 32.** "Now we're going to make the similar forms as last time, only we're going to use the opposite colors."

8 As your child observes, repeat the sequence from the previous lesson; this time use the red crayon for the bars of color and the blue crayon for the waves. Also, this time start with the wavy blue line. Use the medium side of a block crayon to measure for the white spaces.

9 Close your books, put away your materials, and stand for the closing verse.

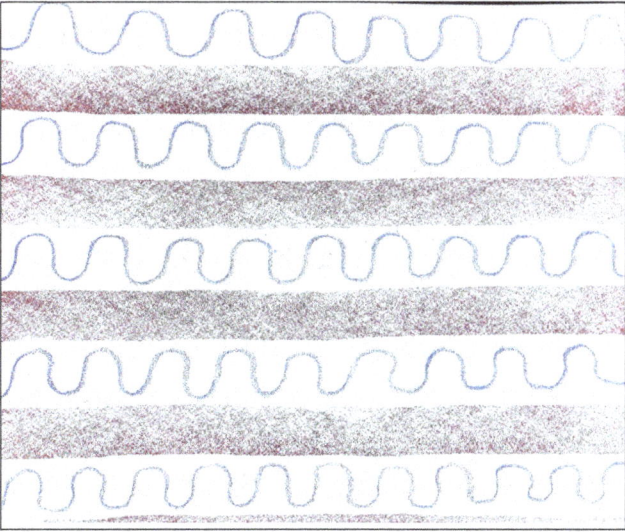

SKILLS 3-5

Focus: Ssubitizing, addition/subtraction factors of 12.

Materials:
- Dot cards
- ML Book #1
- Block and stick crayons
- Large colored pencils
- Counting stones

Lesson Content:

1 Stand to say the opening verse.

2 Bring out the dot cards and follow the exercises listed in **Skills Block Three: Lesson 1**.

3 Turn to **page 33.** Talk with your child about the little squirrels. Have a conversation about their physical bodies and their behaviors. Squirrels have little heads with little pointed noses and bright black eyes. They skitter along the ground, up tree trunks, across branches. They don't stay still for very long, or very often; but when they do, they rest on their strong haunches. Squirrels seem to balance themselves, whether moving or still, with their long, fluffy tails. They have cute little front paws that almost look like hands, and they have little ears and sharp little teeth. Squirrels eat nuts and seeds, and they are always looking for food.

4 [This description is of course for those who live where squirrels live. If you don't, think of another creature that gathers nuts and seeds.] In the autumn in eastern North America, squirrels are busy, busy, busy, gathering acorns to hide away for their winter food. They hide the acorns in hollow trees and in the holes they dig in the ground. We're going to draw two little squirrels and their trees.

5 First, draw the earth. Now draw the trees, then draw the squirrels. [See the illustration.] Now draw **twelve** acorns, all in a row, on the ground between them. In the center, draw a "number tree." This is actually a tall "T." On the cross bar at the top, draw a 12. Now you have a number tree for 12, with space on each side for the factors.

6 "There are 12 acorns on the ground. If this squirrel picks up 12, how many are left for the squirrel in the other tree?" Your child will say "none are left."

7 Continue to describe the first squirrel as picking up a certain number of acorns: 11, 10, 9, 8, 7, 6, 5, 4, 3, 2, 1, 0. In each case, how many are there for the other squirrel? 1, 2, 3, 4, 5, 6, 7, 8, 9, 10, 11, 12. When you have drawn each set of two numbers, underline the set with a straight line.

8 Now ask your child, "What do you notice about our numbers?" If your child does not notice that, as one column diminishes, the other increases, give them hints to help them see it.

9 Now you can work with the factors. Going down the "tree," 12 = 12 + 0; 12 = 11 + 1; 12 = 10 + 2; etc.

10 Working with subtraction and again going down the "tree," 12 - 12 = 0; 12 - 11 = 1; 12 - 10 = 2.

11 This is a very visual representation of the addition and subtraction factors of 12. It also indicates that these two operations are opposites.

12 Close your books, put away your materials, and stand for the closing verse.

SKILLS 3-6

Focus: Counting to 36 by twos and to 100 by fives and tens; addition and subtraction factors of teng.

Materials:
- Jump rope
- ML Book #1
- Stick crayons
- Large colored pencils
- Jar lid or glass, about 3" [7cm.] in diameter
- Counting stones

Lesson Content:

1 Stand to say the opening verse.

2 Practice counting, as above, while jumping rope.

3 Sit, and open the ML book to **page 33.** "In our previous lesson, we drew this page. Can you tell me what it means?" Wait for your child's response, and use it to start a review of the previous lesson.

4 "We drew one way to show how numbers help one another make TWELVE. There are other ways to show this. Today we're going to show how numbers help one another make the number TEN."

5 Turn to **page 34.** In the center of the top of the page, write the number **10.** Beneath this, draw three vertical and parallel lines, about ½" apart, to the bottom of the page.

6 Starting in the upper left hand corner of the page, trace around the jar lid or glass to create a circle. You will need **eleven** such circles on this page.

7 Divide each circle in half with a straight vertical line.

8 "Now, suppose I give you 5 marbles. Let's draw those 5 marbles on one side of the first circle we drew. How many more marbles will we need in order to make 10?"

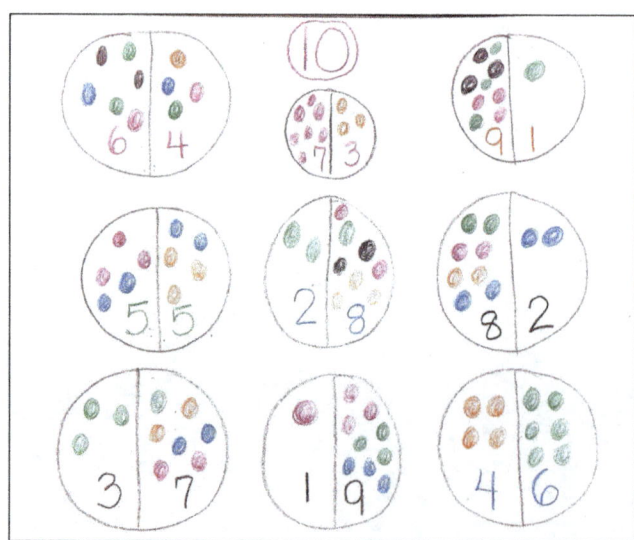

9 Continue to draw marbles on one half of each circle, one circle at a time, and ask your child to tell you how many more marbles you will need to make ten. There are eleven combinations of numbers that will make ten:

5/5	6/4	7/3	4/6	9/1	1/9
10/0	8/2	3/7	0/10	2/8	

10 Children really enjoy figuring out what the combinations are. Once you've completed all eleven circles, you're ready to fill in the vertical columns you drew at the beginning of the exercise, starting with 10/0 and ending with 0/10. You draw the 10/0 and ask your child to tell you which combination comes next. [It is 9/1.]

11 Practice counting backwards from 10 to 0, and forwards from 0 to 10, pointing to each number as you say its name.

12 Put away your materials and stand to say the closing verse.

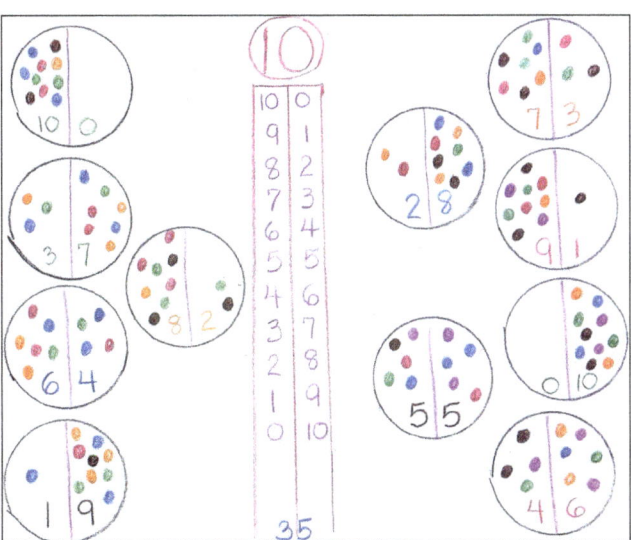

SKILLS 3-7

Focus: Division factors for 12.

Materials:
- Plasticine of two colors—one color for "bird baths" and another color for the birds
- ML Book #1
- Colored pencils

Lesson Content:

1. Stand and say the verse to begin the lesson.

2. With "bird bath" colors, model 12 simple "bird baths."

3. With the "bird" color, model 12 little "birds."

4. "Today we're going to talk about our friend Little Divide. What do you remember about Little Divide?" [Your child might remember that Divide is red, for example.]

5. "Little Divide wants to help the birds find bird baths. They've been flying for a long time and they really want to play in the water. Here are three bird baths. How many birds need to use each bath?" Demonstrate this using the models you've made. Help your child to deliberately count out the birds, placing each in a bath one at a time.

6. "There are three bird baths and twelve birds. So we say '12 birds divided by 3 baths means there will be 4 birds in each bath. 12 divided by 3 = 4.'"

7. "Suppose there are four bird baths. How many birds will play in each bird bath?" Ask your child to figure this out using the models they've made.

8. Continue with this exercise, using the models, for the "arithmetic stories" for:

   12 divided by 2   12 divided by 1
   12 divided by 6   12 divided by 2

9. Open your books to **page 35.** Using block crayons to measure, divide your page into 8 blocks. Start by drawing a vertical line down the middle of the page from top to bottom. Now place all 12 block crayons, on the medium side, down the side of the page. Under each three crayons, place a dot; then remove the crayons.

FIRST GRADE ARITHMETIC—TEACHER'S GUIDE 181

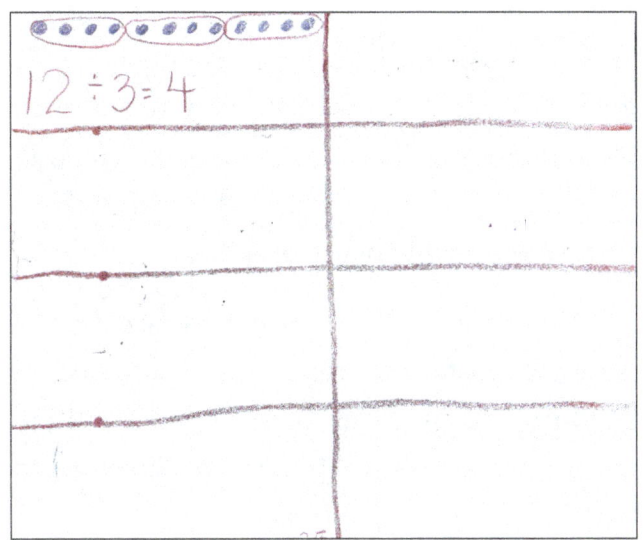
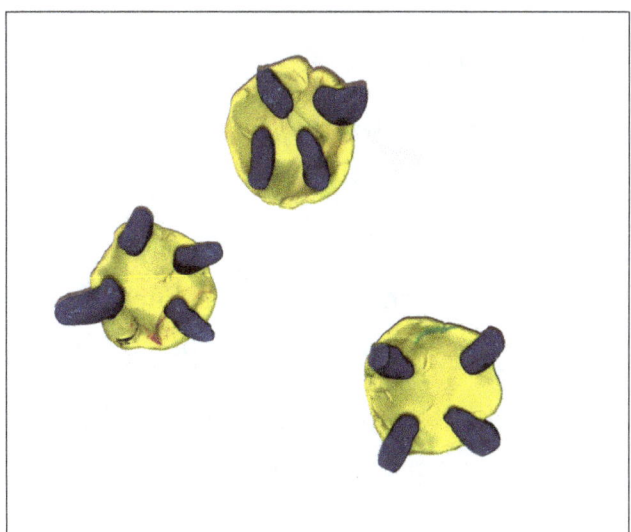

10 Now draw horizontal lines across the page, starting with each dot. You have divided your page into 8 blocks.

11 In each block, write one of the exercises noted above. Encourage your child to use their models in reviewing the exercises.

12 When you have written all 6 of the exercises, ask your child to read them aloud. [You may color in the extra two blocks, leave them blank, or fill them with birds!]

13 Close your books, put away your materials, and say the closing verse.

SKILLS 3-8

Focus: Review of multiplication.

Materials:
- ML Book #1
- Colored pencils
- Plasticine models from previous lesson

Lesson Content:

1 Stand to say the opening verse.

2 "Let's count our bird baths and birds." Ask your child to count each set of models out loud. There should be 12 of each.

3 "Our little birds have been traveling for so long that they have made friends with one another. Sometimes there is a group of three friends, sometimes four or another number. Of course, when they have made friends, they want to play together in the same bath. The first time they come to the baths, they are in groups of three. How many baths will they need?"

4 Continue with the story by grouping the birds in fours, sixes, twos, ones, and as one large group of 12. Instruct your child to use their models to figure out how many baths will be needed in each case.

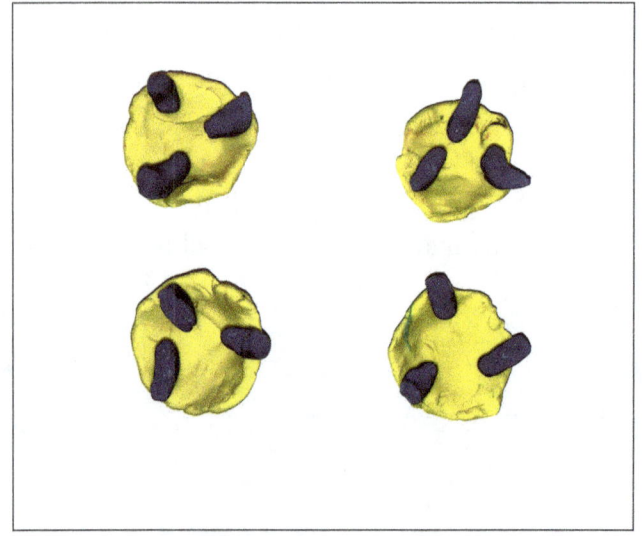

5 Open your books to **page 36.** Create 8 blocks on the page by following the directions in Lesson 7.

6 Now review and write each exercise:

12 = 4 x 3   12 = 3 x 4
12 = 2 x 6   12 = 6 x 2
12 = 1 x 12   12 = 12 x 1

[You will again have two extra blocks. Color them in, leave them blank, or fill them with birds.]

7 Close your books, put away your materials, and stand to say the closing verse.

SKILLS CLASS BLOCK FOUR

Review of the Basic Concepts

AS THE LAST ARITHMETIC BLOCK OF THE YEAR, these eight lessons will lead you through a review of the basic concepts introduced in the previous months. Please remember that each year of the Waldorf curriculum follows a dual path: introduce new material in the spring, and review it in the following fall. So, the material you have studied this year will be reviewed in the first arithmetic blocks of the Second Grade; and those concepts will be expanded upon in the spring of that year. We do not expect children to master the material as we introduce it. New concepts require a period of "sleep." During this time, although a child does not consciously engage with the new concepts, they are being incorporated into what the child already knows.

We begin these lessons with a new activity–that of writing with the foot. Try it yourself! This is usually an enjoyable challenge for children, and it is a subtle way to strengthen one's handwriting skills. As the child literally writes with their foot, their eye, arm, and hand are following the motions.

We use the story of "The Twelve Dancing Princesses" to illustrate the times table for the twos. This provides an opportunity to review "times" and to think of ways we might illustrate other tables. Reviewing one operation leads into a review of all four as we model the operational signs out of beeswax. Creating a rope of beads leads to reviewing counting by fives and tens as well as counting by ones to 100.

We also review subitizing with our old friends the dice. Now we learn how to "count on," again using the dice and subitizing skills. If the first die shows a 4 and the second die shows a 6, we count from the 4 or the 6 rather than starting at one.

Now we learn a new skill, that of comparing numbers. When we see two numbers, which is the larger and which is the smaller? In our final lesson of the block, we review counting by fives and tens, comparing numbers, and "counting on." These reviews give us an opportunity to point out what the child has learned in this first year of arithmetic and how well they are doing. Look back to what the child knew at the beginning of the year. Let the child know how far they have come!

Focus: Multiplying by two.

Materials:
- ML Book #1
- Block and stick crayons
- Story [below]: "The Twelve Dancing Princesses"
- Dot cards
- Large piece [at least 18" x 24"] of heavy paper [You can use a piece of tag board.]

Lesson Content:

1. Stand for the opening verse.

2. Ask your child to remove their shoes and socks. Place the piece of tag board under your child's feet. Ask your child to choose a foot to draw with. [Your child's choice will indicate which foot is dominant.]

3. Place a stick crayon between the big toe and the toe next to it. Ask your child to write the numbers 1 - 6 with their foot. [As most children have no experience with this activity, it will be easier for your child if they start by tracing the numbers. This will require you to prepare the model before class time.]

4. Put away the foot-writing materials and put your child's socks and shoes back on. Instruct your child to sit up straight and tall, hands on the desk.

5. Practice subitizing with the dot cards as you did in earlier lessons.

6. Now tell the story of **The Twelve Dancing Princesses.**

7. Stand for the closing verse or song.

The Twelve Dancing Princesses

Skills Class Four

Grimm's

THERE WAS A KING who had twelve beautiful daughtersand would be the next king. They slept in twelve beds all in one room and when they went to bed, the doors were shut and locked up. However, every morning their shoes were found to be quite worn through as if they had been danced in all night. Nobody could find out how it happened, or where the princesses had been.

So the king made it known to all the land that if any person could discover the secret and find out where it was that the princesses danced in the night, he would have the one he liked best to take as his wife, and would be king after his death. But whoever tried and did not succeed, after three days and nights, they would be put to death.

A king's son soon came. He was well entertained, and in the evening was taken to the chamber next to the one where the princesses lay in their twelve beds. There he was to sit and watch where they went to dance; and, in order that nothing could happen without him hearing it, the door of his chamber was left open. But the king's son soon fell asleep; and when he awoke in the morning he found that the princesses had all been dancing, for the soles of their shoes were full of holes.

The same thing happened the second and third night and so the king ordered him to be killed.

After him came several others; but they all had the same luck, and all lost their lives in the same way.

Now it happened that a soldier, who had been wounded in battle and could fight no longer, passed through the country where this king reigned, and as he was travelling through a wood, he met an old woman, who asked him where he was going.

"I hardly know where I am going, or what I had better do," said the soldier; "but I think I would like to find out where it is that the princesses dance, and then in time I might be a king."

"Well," said the old woman, "that is not a very hard task: only take care not to drink any of the wine which one of the princesses will bring to you in the evening; and as soon as she leaves you pretend to be fast asleep."

Then she gave him a cloak, and said, 'As soon as you put that on you will become invisible, and you will then be able to follow the princesses wherever they go.' When the soldier heard all this good advice, he was determined to try his luck, so he went to the king, and said he was willing to undertake the task.

He was as well received as the others had been, and the king ordered fine royal robes to be given him; and when the evening came he was led to the outer chamber.

Just as he was going to lie down, the eldest of the princesses brought him a cup of wine; but the soldier threw it all away secretly, taking care not to drink a drop. Then he laid himself down on his bed, and in a little while began to snore very loudly as if he was fast asleep.

When the twelve princesses heard this they laughed heartily; and the eldest said, "This fellow too might have done a wiser thing than lose his life in this way!" Then they rose and opened their drawers and boxes, and took out all their fine clothes, and dressed themselves at the mirror, and skipped about as if they were eager to begin dancing.

But the youngest said, "I don't know why it is, but while you are so happy I feel very uneasy; I am sure some mischance will befall us."

"You silly person," said the eldest, "you are always afraid; have you forgotten how many kings' sons have already watched in vain? And as for this soldier, even if I had not given him his sleeping draught, he would have slept soundly enough."

When they were all ready, they went and looked at the soldier; but he snored on, and did not stir hand or foot: so they thought they were quite safe.

Then the eldest went up to her own bed and clapped her hands, and the bed sank into the floor and a trap-door flew open. The soldier saw them going down through the trap-door one after another, the eldest leading the way; and thinking he had no time to lose, he jumped up, put on the cloak which the old woman had given him, and followed them.

However, in the middle of the stairs he trod on the gown of the youngest princess, and she cried out to her sisters, "All is not right; someone took hold of my gown."

"You silly creature!" said the eldest, "it is nothing but a nail in the wall.'

Down they all went, and at the bottom they found themselves in a most delightful grove of trees; and the leaves were all of silver, and glittered and sparkled beautifully. The soldier wished to take away some token of the place; so he broke off a little branch, and there came a loud noise from the tree. Then the youngest daughter said again, "I am sure all is not right -- did not you hear that noise? That never happened before."

But the eldest said, "It is only our princes, who are shouting for joy at our approach."

They came to another grove of trees, where all the leaves were of gold; and afterwards to a third, where the leaves were all glittering diamonds. And the soldier broke a branch from each; and every time there was a loud noise, which made the youngest sister tremble with fear. But the eldest still said it was only the princes, who were crying for joy.

They went on till they came to a great lake; and at the side of the lake there lay twelve little boats with twelve handsome princes in them, who seemed to be waiting there for the princesses.

One of the princesses went into each boat, and the soldier stepped into the same boat as the youngest. As they were rowing over the lake, the prince who was in the boat with the youngest princess and the soldier said, "I do not know why it is, but though I am rowing with all my might we do not get on so fast as usual, and I am quite tired: the boat seems very heavy today."

"It is only the heat of the weather," said the princess, "I am very warm, too."

On the other side of the lake stood a fine, illuminated castle from which came the merry music of horns and trumpets. There they all landed, and went into the castle, and each prince danced with his princess; and the soldier, who was still invisible, danced with them too. When any of the princesses had a cup of wine set by her, he drank it all up, so that when she put the cup to her mouth it was empty. At this, too, the youngest sister was terribly frightened, but the eldest always silenced her.

They danced on till three o'clock in the morning, and then all their shoes were worn out, so that they were obliged to leave. The princes rowed them back again over the lake (but this time the soldier placed himself in the boat with the eldest princess); and on the opposite shore they took leave of each other, the princesses promising to come again the next night.

When they came to the stairs, the soldier ran on before the princesses, and laid himself down. And as the twelve, tired sisters slowly came up, they heard him snoring in his bed and they said, "Now all is quite safe."' Then they undressed themselves, put away their fine clothes, pulled off their shoes, and went to bed.

In the morning the soldier said nothing about what had happened, but determined to see more of this strange adventure, and went again on the second and third nights. Everything happened just as before: the princesses danced till their shoes were worn to pieces, and then returned home. On the third night the soldier carried away one of the golden cups as a token of where he had been.

As soon as the time came when he was to declare the secret, he was taken before the king with the three branches and the golden cup; and the twelve princesses stood listening behind the door to hear what he would say.

The king asked him. "Where do my twelve daughters dance at night?'"

The soldier answered, "With twelve princes in a castle underground." And then he told the king all that had happened, and showed him the three branches and the golden cup which he had brought with him.

The king called for the princesses, and asked them whether what the soldier said was true and when they saw that they were discovered, and that it was of no use to deny what had happened, they confessed it all.

So the king asked the soldier which of the princesses he would choose for his wife; and he answered, "I am not very young, so I will have the eldest.'" –- and they were married that very day, and they were chosen to rule the kingdom when the king had died.

Focus: Illustrate story: *The Twelve Dancing Princesses*.

Materials:
- ML Book #1
- Block and stick crayons
- Large colored pencils

Lesson Content:

1 Stand for the opening verse.

2 Review the story from the previous lesson. Encourage your child to tell as much of the story as they remember. This review should not last longer than ten minutes.

3 Turn to **page 38** in your book. [Skip page 37, as lessons 2 and 3 need to be drawn opposite one another.]

4 Instruct your child to draw twelve princesses. The princesses and their clothing, etc., can be any color; but they need to be small enough to all fit on one page. And each princess must have two shoes!

5 This will take the entire lesson. Close your books, put away your materials, and stand for the closing verse.

Focus: Multiplying by two.

Materials:
- ML Book #1
- Block and stick crayons
- Large colored pencils
- Play ball

Lesson Content:

1. Stand and say the verse to begin the lesson.

2. Stand with the play ball and bounce as you skip count by twos. Bounce on the number and be silent when you catch the ball.

3. Turn to **page 39.** Using the medium side of a red and a yellow block crayon, draw alternating bars of color from the top of the page to the bottom of the page. You should have 12 bars of color: 6 bars of red and 6 bars of yellow.

4. On the top bar of color, next to the spiral binding, draw one small dancing princess with two shoes.

5. Next to her, draw a **1**. All the way across the page, on the right margin of the top bar of color, draw a **2**. On the left side of the **2**, draw a pair of shoes. Explain to your child that **one** princess would have **two** shoes.

6. On the next bar of color, under the **1**, draw a **2**. Ask your child: "How many shoes would **two** princesses have?" Under the **2** on the right margin, draw a **4**. On the left side of the **4**, draw two sets of two shoes each.

7. Continue down the page, with numbers 1 - 12 on the left and 2 - 24 on the right.

8. Now "read" the page: you say, "One princess has…" and your child responds, "Two shoes." Continue until you have finished with, "Twelve princesses have….24 shoes."

9. Close your books, put away your materials, and stand for the closing verse.

Focus: Modeling and review of the symbols for the four operations.

Materials:
- Dot cards
- ½ stick of each color beeswax: yellow, red, green, blue
- Number path [for your child to use when working with plus and minus]

Lesson Content:

1 Stand to say the opening verse.

2 Take out your green beeswax and prepare to warm it. Depending on the time of year and the temperature in your school space, it might be soft enough to model. If not, you can put it under your arms or even sit on it. Keep it in your "oven" until finished with the subitizing activity.

3 Using the dot cards, practice subitizing with your child. Start out by showing each card for a few seconds. Gradually increase the speed with which you take away the card. Remember that your goal is to help your child recognize the number of dots automatically, without counting.

4 Take out the green beeswax. [At the same time, place the yellow beeswax in your "oven."] Mold it into an egg shape, and then model it into the little green figure shown in the photograph. Choose a different color of beeswax to create the "+" sign. While modeling, talk about addition: "Farmer Jo/e has 11 eggs altogether. There are 5 in one nest; how many are in the other? So eleven is the same as five plus six." Use the largest numbers your child is comfortable with.

Examples:
15 = 11 + ?
28 = 22 + ?
19 = 12 + ?

5 When you've finished modeling "Little Plus," take out the yellow beeswax and mold it into an egg shape. [At the same time, place the red beeswax in your "oven."] Then model it into the yellow figure as shown in the photograph. Use a different color of beeswax to form the "x" symbol. While modeling, talk about multiplication/times: "Farmer Jo/e has four neighbors, and she wants to give three eggs to each. How many eggs will she need? 4 x 3 = 12."

Examples: 2 x 6   3 x 5   4 x 4   11 x 2

6 When you've finished modeling "Little Times," take out the red beeswax and mold it into an egg shape. [At the same time, place the blue beeswax in your "oven."] Model the red figure in the photograph, using a different color of beeswax to create the sign for division. While modeling, talk about division: "Farmer Jo/e has 12 eggs, and she wants to give 4 eggs to each neighbor. How many neighbors will she be able to give 4 eggs? 12 divided by 4 = 3.

Examples:
14 divided by 2
10 divided by 5
12 divided by 3

7 When you've finished modeling "Little Divide," take out the blue beeswax and mold it into an egg shape. Model it into the blue figure as shown in the photograph. Use a different color of beeswax to form the "–" symbol. While modeling, talk about "Little **MInus:**" "Farmer Joe had 10 eggs, but the rat ate 6. How many were left?" 10 – 6 = 4.

Examples: 15 - 6   18 - 7   11 - 5

8 When you have finished modeling all four figures, review the signs and what they mean.

9 Place your figures in a safe place and stand to say the closing verse.

First Grade Arithmetic—Teacher's Guide

Focus: Counting by 5 and by 10.

Materials:
- Jump rope or play ball
- Strong cord 6' in length
- 50 red wooden beads
- 50 white wooden beads
  [The beads can be any two colors. Make sure your cord will thread through them.]

SKILLS 4-5

Lesson Content:

1. Stand to say the opening verse.

2. Count by 5s to 100 while jumping rope or bouncing the play ball.

3. Count by 10s to 100 [or beyond] while jumping rope or bouncing the play ball.

4. Tie a large knot in one end of the cord.

5. String the beads, five red, then five white, until you have used all 100 beads.

6. Count by 5s and then 10s using your number cord.

7. Put away your materials and stand for the closing verse.

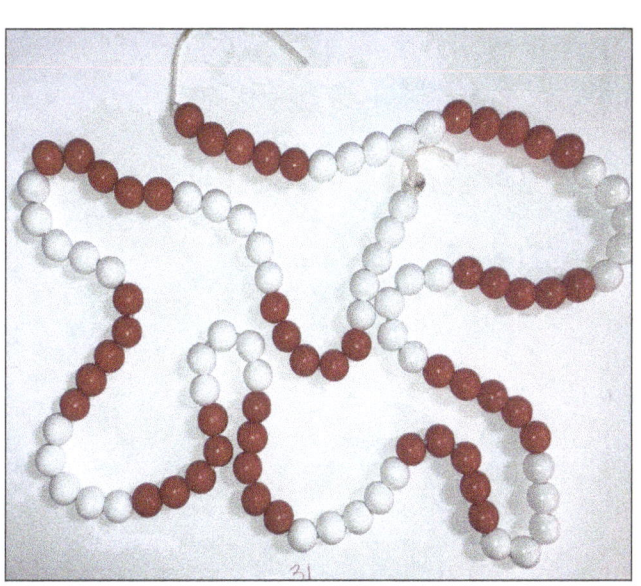

Focus: Subitizing and comparing numbers.

Materials:
- Two dice
- Dot cards
- Chalkboard, chalk, and eraser

Lesson Content:

1 Stand to say the opening verse.

2 Review yesterday's lesson. Write several number pairs on the chalkboard and ask your child to fill in the correct sign for comparing them.

3 Ask your child to sit down again. Hold up the dot cards, one at a time, and ask your child to name the quantity of dots on each card as fast as they can.

4 Put the dot cards away and take out the dice. Roll one die on your child's desk and ask them to name the number of dots on the face of the die.

5 Ask your child to roll the die and tell you the number of dots. Do this several times.

6 Now roll one die and ask your child to name the number. Leaving that die in place, roll the other one. "How many dots are there now?" The point is for your child to already know the number of dots on one die and to count on from that total to name the sum of the two dice together. This is a practical demonstration of why we learn to subitize.

7 Repeat this exercise several times.

8 Put away your materials and stand to say the closing verse.

Focus: More than/less than.

Materials:
- ML book #1
- Block and stick crayons
- 24 counting stones
- Large colored pencils
- Four beeswax figures representing symbols of Arithmetic

Lesson Content:

1. Stand to say the opening verse.

2. "We are very close to the end of First Grade, and by now you know many things. For example, you know that in Arithmetic, we often use *symbols* instead of *words*." Hold up the beeswax figures, one at a time. With each figure, point to the sign itself and say, "This is a *symbol* that we have learned in Arithmetic. What does this symbol mean in words?"

3. When you have reviewed all four symbols, say, "Today we are going to learn two new symbols. Sometimes when we are working with two numbers, we need to show which number is larger or which number is smaller. We use those symbols instead of writing all the words in this arithmetic sentence: 'Seven is larger than four.' This is how we write that arithmetic sentence in arithmetic language: 7 ➤ 4. See how much simpler that is?"

4. Count out the counting stones to make sure there are 24. Now separate the stones into two random piles and count the stones in each pile. "Which pile has more stones?"

5. Repeat this sequence. Now ask, "Which pile has less?"

6. Repeat this sequence until you are certain your child understands these concepts. [You might not need to repeat it at all.]

7. Open your books to **page 40.** Stack four block crayons on the medium side down the upper left-hand margin of the page. Draw a light dash under the bottom crayon and remove the block crayons. Now draw a straight line, with a blue stick crayon, across the page.

8. Using this line as a guide, use your blue block crayon to draw a bar of color across the page. Under the blue bar, draw a green bar. Continue to draw alternate bars of color until you reach the bottom of the

page. You should have 7 or 8 bars of alternating colors.

9. On the blue line, draw two alligators as shown in the photograph. Be sure to draw their hungry mouths wide open! Draw a bright, cheerful sun in the middle of the page above them.

10. "Sometimes we use pictures to help us remember things. In this picture, when we look at the alligator's hungry mouth, it is easy to remember that if the alligator is hungry, that means it has less! So if I write these two numbers, which one is hungry?" Write 5 and 8 on the chalkboard, leaving a space between them. Your child should respond that "5" is hungry, as it is the smaller number. "So we write the sign for "less than," because it looks like a hungry alligator.

11. Now draw the numbers again, but this time write 8 and 5. "Which number is hungry this time? Yes! Five is still hungry, so we draw *this* sign because it looks like a hungry alligator."

12. Place 4 block crayons on the medium side, end to end, on the top bar of color. Make a little mark at the end of the line. With your pencil, draw a straight vertical line to the bottom of the page. You've now created two columns for your work on this page.

13. Write on the chalkboard 4 and 8, with a space between. Ask your child to copy these numbers in their book on the top line of the left-hand column. Ask your child which is the smaller number; and instruct them to write the correct sign in the space between.

14. Continue to write numbers on the chalkboard for your child to copy into their book. If your child seems to be understanding these concepts, you can write several number pairs on the chalkboard. Be sure to mix up the values of the numbers so that they are not in any perceivable pattern of "more than" or "less than."

15. When your child has completed the two columns of number pairs, say, "When we work with numbers like this, we say that we are *comparing* numbers." Talk about what it means to compare two things to one another. For example, you might have two chairs or two tables, one of which is larger–and one of which is smaller–than the other. "In arithmetic, we are working with numbers, so we need this way to *compare* the numbers."

16. Close your books, put away your materials, and stand for the closing verse.

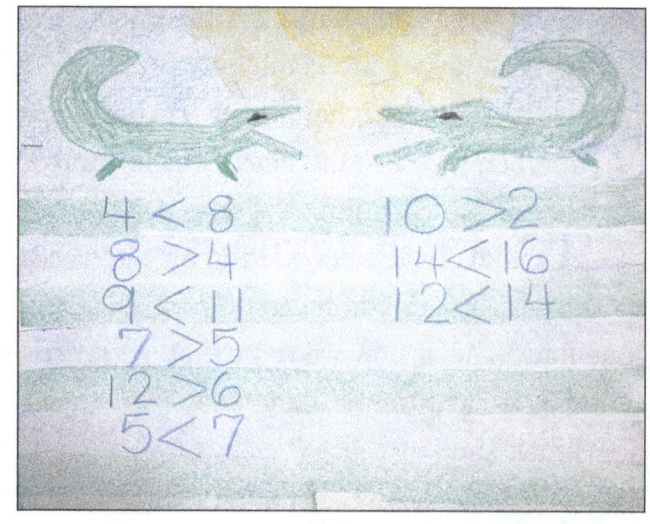

Focus: Review of Skills lessons in this Block..

Materials:
- Play ball
- 100's rope
- Two dice

Lesson Content:

1. Stand to say the opening verse.

2. With the play ball, count by twos as high as you can. Great if you can count to 24 or 36; amazing if you can get to 100; magnificent if you can keep going!

3. With the 100's rope, count to 100 by 10s and then to 100 by 5s. Can you start over on the rope and count by tens or fives beyond 100?

4. Write several number pairs on the chalkboard. Ask your child to write the correct signs for comparing the two numbers in each pair.

5. Ask your child to roll one die and shout out the number of dots. Then add another die. Practice subitizing and "counting on" as in the previous lesson.

6. Put away your materials and stand to say the closing verse.

APPENDICES

"Arithmetic" rather than "Math" • 202

Bibliography of Books and Articles • 203

Children's Art Developmentr • 208

Drawing With Children • 210

Making Yoir Own Materials • 212

Pencil Grip and Posture • 214

Poems and Songs • 215

Samples of Title Pages • 233

Supplemental Activities • 235

Supplemental Sources for Activities • 237

The Temperaments • 238

"What if my child doesn't seem to be responding? • 248"

"Arithmetic" rather than "Math"

As a side note, you might be wondering why we refer to "Arithmetic" rather than to "math." "Math," after all, is the more familiar term. In the "spiral" curriculum as currently taught, it might even be the accurate term, as in schools which have adopted that type of curriculum, several branches of Mathematics are taught from the earliest grades. In the early grades of the Waldorf school, however, we teach Arithmetic, which is the foundational branch on the broad-reaching tree of Mathematics. Mathematics as a subject incorporates the study of: arithmetic, algebra, geometry, trigonometry, statistics & probability, and calculus, as well as several subjects which fall into the category of "Higher Math." Because the Waldorf curriculum is a developmental curriculum, we start with "pure" arithmetic in Grades 1-4 or 5, and move into a formal study of algebra and geometry in middle school. "Arithmetic" includes the operations of addition, multiplication, subtraction, and division of whole numbers, common fractions, and decimal fractions. By firmly grounding the children in the operations of Arithmetic before proceeding to more complex mathematics, we believe that children develop the confidence that comes with understanding and the enthusiasm that comes with discovery. Our approach proceeds logically, laying a foundation of concrete operations and then gradually moving into the more abstract operations.

A Bibliography of Books and Articles

Consulted in Writing this Manual

Books on Teaching Arithmetic:

Belenson, Mel: *Working Material for the Class Teacher,* published as a manuscript by the Pedagogical Section Council and the Association of Waldorf Schools of North America, 1996.

Boaler, Jo: *Mathematical Mindsets,* Jossey-Bass, copyright 2016 by Jo Boaler.

Boaler, Jo: *What's Math Got to Do With It?* Penguin Books, copyright 2015 by Jo Boaler.

Fabrie, Nettie, Wim Gottenbos, and Jamie York: *Making Math Meaningful,* Jamie York Press, copyright 2009 by Fabrie, Gottenbos, and York.

Gottgens, Else: *Waldorf Education in Practice,* Outskirts Press Inc., copyright 2011 by Else Gottgens.

Harrer, Dorothy: *Math Lessons for Elementary Grades,* the Association of Waldorf Schools of North America, copyright 1985 by the Association of Waldorf Schools of North America.

Newton, Nicki: *Guided Math Lessons in First Grade,* Routledge, copyright 2022 by Taylor and Francis.

Parrish, Sherry: *Number Talks: Helping Children Build Mental Math and Computation Strategies*, Math Solutions, copyright 2014 by Sherry Parrish.

Schuberth, Ernst: *Teaching Mathematics for First and Second Grades in Waldorf Schools,* Rudolf Steiner College Press, copyright 1999 by Ernst Schuberth.

Shumway, Jessica F.: *Number Sense Routines: Building Numerical Literacy Every Day in Grades K - 3,* Stenhouse Publishers, copyright 2011 by Jessica Shumway.

Sousa, David A.: *How the Brain Learns Mathematics,* Sage Publications Ltd., copyright 2015 by David A. Sousa.

Sousa, David A.: *How the Brain Learns,* Sage Publications Ltd., copyright 2017 by David A. Sousa.

Steiner, Rudolf: *Faculty Meetings with Rudolf Steiner,* copyright Anthroposophic Press, 1998.

The Spiritual Ground of Education, copyright Anthroposophic Press, 2004.

The Renewal of Education, published by permission of the Rudolf Steiner Nachlassverwaltung, Dornach, Switzerland, 1989.

Discussions with Teachers, copyright Rudolf Steiner Press, London, 1967.

Practical Advice to Teachers, copyright Rudolf Steiner Press, London, 1976.

The Kingdom of Childhood, copyright Rudolf Steiner Press, London, 1982.

A Modern Art of Education, copyright by Garber Communications, 1989.

Stockmeyer, E.A. Karl: *Rudolf Steiner's Curriculum for Waldorf Schools,* copyright 1985 by the Steiner Schools Fellowship.

Von Baravalle, Hermann: *The Teaching of Arithmetic and The Waldorf School Plan,* Waldorf School Monographs, copyright 1967 by Hermann von Baravalle.

Articles on Teaching Arithmetic:

Banerji, Olina: *A District Is Making a Huge Bet on One Math Curriculum to Improve Achievement.* www.edweek.org, June 27, 2024.

Banerji, Olina: *Word Problems Get a Bad Rap in Math Class. Here's How to Get Them Right.* www.edweek.org, October 15, 2024.

Devitt, Rebecca: *Mastery vs Spiral-Based Learning: What Works Best?* https://Howdoihomeschool.com, December 14, 2022.

Education Week Spotlight 2023: *Improving Math Fluency.*
 Sawchuk, Stephen: *What Is Math 'Fact Fluency,' and How Does it Develop?*
 Schwartz, Sarah: *Why Word Problems Are Such a Struggle for Students–and What Teachers Can Do*
 Sawchuk, Stephen: *Kids Need to Know Their Math Facts. What Schools Can Do to Help*
 Sparks, Sarah: *These Early Math Supports Translated to Gains Later on for Vulnerable Students.*
 Vaughn, Viveka: Math Trauma Is Real. Here's How You Can Prevent It.

Ferlazzo, Larry: *Response: Ways to Teach Math Besides 'Drill the Skill,'* www.edweek.org, October 27, 2014.

Klein, Alyson: *Teachers Are Evenly Divided on the Best Way to Teach Math.* www.edweek.org, May 30, 2024.

Larrison, Abigail, Alan Daly, and Carol VanVooren: *Twenty Years and Counting: A Look at Waldorf in the Public Sector Using Online Sources.* Current Issues in Education, Vol. 15, No. 3, October 5, 2012.

Lord, Becky: *Five Instructional Shifts to Support Students with Dyscalculia,* becky@lordmath.com.

Lord, Becky: *My #1 Recommendation: Use Dot Cards,* becky@lordmath.com, November 12, 2024.

Lord, Becky: *What Is Structured Numeracy?* becky@lordmath.com., January 22, 2025.

National Council of Teachers of Mathematics: *Mathematical Proficiency: the Five Strands.*

Sawchuk, Stephen: *Do Timed Tasks Really Worsen Math Anxiety?* www.edweek.org, August 7, 2024.

Schwartz, Sarah: *California Adopts controversial New Math Framework. Here's What's in It.* www.edweek.org, July 12, 2023.

Schwartz, Sarah: *Math Teachers and Math Ed. Professors Don't See Eye to Eye on Best Practices.* www.edweek.org, October 13, 2023.

Schwartz, Sarah: *Schools Prioritize Reading Intervention. But What About Math?* www.edweek.org, February 23, 2024.

Schwartz, Sarah: *Six Components of Effective Math Games.* www.edweek.org, November 13, 2023.

Schwartz, Sarah: *Why Does Fact Fluency Matter in Math? Four Educators Offer Answers.* www.edweek.org, December 7, 2023.

Schwartz, Sarah: *Why Word Problems Are Such a Struggle for Students–And What Teachers Can Do,* www.edweek.org, May 01, 2023.

Solis, Vaness & Stephen Sawchuk: *How Addition Fluency Develops: A Visual Explainer,* www.edweek.org, May 11, 2023.

Movement

Goddard Blythe, Sally: Movement: *Your Child's First Language,* Hawthorn Press, copyright 2018 by Sally Goddard Blythe and Michael Lazarev.

Goddard Blythe, Sally: *The Well Balanced Child: Movement and Early Learning,* Hawthorn Press.

Hannaford, Carla: Smart Moves: *Why Learning Is Not All in Your Head,* Great Ocean Publishers, copyright 1995 by Carla Hannaford.

Andersen, Henning: *Active Arithmetic!* Association of Waldorf Schools of North America, copyright 1995.

Tunkey, Jeff: *Educating for Balance and Resilience,* Bell Pond Books, copyright 2020 by Jeff Tunkey.

Nash-Wortham, Mary, & Jean Hunt: *Take Time,* The Robinswood Press, copyright 1990 by Nash-Wortham and Hunt.

Fairy Tales and Stories

Phelps, Ethel Johnston: *The Maid of the North,* Holt, Rinehart, and Winston. Copyright 1981 by Ethel Johnston Phelps.

Pantheon Books: *The Complete Grimm's Fairy Tales,* copyright 1972 by Random House, Inc.

Form Drawing

Niederhauser, Hans R. and Margaret Frohlich: *Form Drawing,* Mercury Press, copyright 1984 by Margaret Frohlich.

James, Van: *Language of the Line: A Reinvented Artform of the Waldorf Schools,* Lilipoh, Summer 2005.

Froebe, Carl: Drawing: *From First Grade to High School,* Education as an Art, Vol. 19, Spring 1959.

Kirchner, Hermann: *Dynamic Drawing: Its Therapeutic Aspect*, Mercury Press, translation copyright 1977 by Margarete Frohlich.

Shubert, E. & Embrey-Stine, L.: *Form Drawing Grades One Through Four,* Rudolf Steiner College Press.

Ogletree, Earl J.: *Geometric Form Drawing: A Perceptual-Motor Approach to Preventive Remediation [The Steiner Approach],* The Journal of Special Education, Vol. 9, No. 3, 1975.

Jarman, R.A.: *From Beauty to Truth in Mathematics.*

Sources for Traditional Children's Songs and Fingerplays

Glazer, Tom: *Eye Winker, Tom Tiker, Chin Chopper,* Doubleday and Co., copyright 1973 by Tom Glazer.

Langstaff, Nancy & John: *Sally Go Round the Moon,* Revels Publications, copyright 1986 by Nancy and John Langstaff.

Hart, Jane: *Singing Bee! A Collection of Favorite Children's Songs,* Lothrop, Lee & Shepard Books, copyright 1982 by Jane Hart.

Anything by John Feierabend.

All of the "We Sing" books.

Any collection of Mother Goose rhymes.

Von Heider, Molly: *Looking Forward,* Hawthorn Press, copyright 1995 by Molly von Heider.

Children's Art Development

The Waldorf approach to every subject is artistic. This means that you will be telling stories, singing songs, reciting poetry, and drawing with your child. [See "Lesson Overview."] The very best practice is to learn the stories, songs, and poems by heart so that you will not have to refer to paper while presenting them to your child, and you will be able to do the movements along with your child. Learning a story by heart does not mean you must spend hours memorizing it word for word, but that you know it well enough to tell it in your own words. Sometimes it helps to create mental images of parts of the story in sequence. As you're telling the story, call up those images again and describe what is happening in the scene. You will notice a profound difference in engagement, retention, and connection, if you have taken the time to prepare the story by heart.

Children's drawing is developmental. Do not expect your child to be able to produce a masterpiece of perspective and composition. You can create your own drawings as models or use those provided with each lesson. Regardless, it will be important to practice your own drawing before you show it to your child.

By the time children are six years old, they have often stopped drawing exclusively in lines and "stick figures," and are drawing with blocks of color. They might draw an outline and then "fill it in," or they might draw the block of color by applying close parallel lines with their stick crayons. If your child is still drawing stick figures, it is particularly important for you to guide them into drawing blocks of color.

You'll notice that all story illustrations in this manual are drawn in this way. I have also focused the subject matter on what is most important in the story. The drawing from Lesson A1, for example, depicts only the garden wall, the rose bush, and the path. Modeling simple illustrations helps the child to discern what is most important in the story. In providing the child with a visual imagination of the story, the illustration becomes another support for the child's memory.

One criticism sometimes heard by Waldorf teachers is that "the art all looks alike." When people say this, it simply means that they do not understand that even children who draw spontaneously and beautifully need instruction in developing technique and use of materials. Additionally, although the subject matter in these drawings might be identical, close examination reflects the individuality of the young artists: no two are truly identical.

In the section titled, "Children's Drawings, Step by Step," I outline how to teach drawing with the child's development in mind. I advise you to follow this pattern during class time. After class, if your child so wishes, they can draw their choice of an illustration from the story. You will notice, over time, that the child is incorporating the skills you've taught them into their independent drawings.

Drawing with Children

If given an opportunity and materials, most children draw spontaneously, almost unconsciously. With great enthusiasm, they will draw the same circles or lines over and over again, never getting bored. Children around the world follow the same developmental sequence: random scribbles, circles, circles with "eyes and mouths," circles with "arms and legs," etc. By age 4 or 5, children tell stories with their drawings; by age 5 or 6, they add simple landscapes: a sun in the sky and perhaps a straight line across the bottom for the earth. The people they draw are likewise simple, often stick figures, sometimes without hands or feet. Our job is to model more complex drawings for the child and to gently guide them to a new stage, a new way of transferring what they see onto the page.

We start by watching the child draw and studying the drawing. Is the earth indicated? Is there a tree, a person, a sun, a house? After examining such a spontaneous drawing, Waldorf teachers formally assign "person, house, tree" pictures, to be drawn in that order, to their classes. These drawings indicate what is important to the child as well as their stage of development. It is truly interesting to assign these drawings monthly along with "free" drawings and to study the child's development over the course of the year.

Meanwhile, the teacher creates a drawing for each story and instructs the children to follow it in their own books.

This requires the teacher to plan the drawing ahead of time, thinking about the story and what they want to emphasize in a simple crayon drawing. The story of "The Queen Bee" [ML Block 1, lesson 8] is a good example. Since I was illustrating the spiral form, it was important to create a drawing that focused on the Queen Bee's flight. I also needed to include a sleeping princess and a prince. As you can see in the illustrations for that story, I followed this sequence:

- Horizontal bar of color across the bottom of the page for the floor
- Simple bed with straight horizontal and vertical lines
- Head of the princess with hair and crown
- Body of the princess under a blanket
- Head of the prince with hair and crown
- Neck and shoulders

- Outstretched arm
- Cape
- Legs and feet in boots
- Chandelier
- Wall color

I always start with the head of the person and work down to the feet. When drawing with children, I point to the place on the page where they will put the head for the first few drawings. Then, before they start to draw the person, I ask them, "Where do you think the head will go?" Once I have agreed with their placement, they start drawing. In addition, I do not add facial features to the head, leaving that up to the child. I do explain to them, however, that drawing a face is a real challenge, so I prefer to just imagine it. Sometimes they follow my lead, sometimes they add features. Sometimes they say, "You're right, it is a challenge, and I wish I hadn't done it." If that happens, I remind the child that they will have another chance to make the choice.

I have several goals when following this procedure:

- I want the children to draw full figures rather than stick figures.
- I want the children to use many colors, not just black [which many children will choose, simply because they have been drawing with pens and/or markers].
- I want the children to start with a foundation, be it floor or earth, for the drawing.
- Without explaining or analyzing, I'm modeling proportions and perspective. This work will gain in complexity as the child grows.
- I am also modeling sequence and *planning*–in other words, executive function.

I have, of course, heard parents' concern that "all of the drawings look just alike." Close examination of children's work always demonstrates that every drawing is unique, even though the subject matter is the same. This part of the morning lesson is designed to incorporate drawing instruction; and just like instruction in any skill, everyone starts with basic techniques.

No matter what the child produces, I always find something in the drawing on which I can comment positively. Perhaps I notice the color chosen for clothing or the way the color has been applied with the crayon. Perhaps the house is more complex than the one in a previous drawing, or the child has included a clump of grass or a tiny rabbit hiding in the grass. There will always be something I can point to and smile as I meet the child's eyes.

Making Your Own Materials

Making your own dot cards and number cards

Materials needed:
Sturdy blank index cards
Black permanent marker
  [When finished, laminating these cards is a good idea.]

Although you can easily find dot cards to print and cut out online, it's rather satisfying to make them yourself. [To find them online, just enter "free dot card templates" in your search engine.

There are several designs for dot cards. I've made all of them and use them depending on my goal for the lesson.

1. Draw large dots grouped around the center of the card. Use one card for each number of dots: one, two, three, four, five. These are the cards to use at the beginning of First Grade lessons.
2. Draw large dots in patterns, again grouped around the center of the card. For example:
    a. Two dots can be drawn vertically on one card and horizontally on another.
    b. Three dots can be drawn in a vertical line, a horizontal line, or as triangles.
    c. Four dots can be drawn in a vertical line, a horizontal line, as a square, or as a triangle "plus one."
    d. Five dots can be drawn in a vertical line, a horizontal line, as a "house," or as a square with a dot in the middle.
3. "Doubling" is a helpful skill. To show doubles with the dot cards, draw thin lines horizontally across the center of five cards.
    a. On the first card, draw one dot above the line and one dot below it.
    b. On the second card, draw two dots above the line and one dot below it.
    c. Continue in this manner with 3, 4, and 5 dots.

As your child begins to easily subitize these number patterns, add cards with 6 - 10 dots in arrangements similar to those described above.

Number Cards

50 - 100 sturdy blank index cards

Black permanent marker

[When you finish, laminating these cards is a good idea.]

- Decide whether you want to use half an index card or a full index card for each number.
- Write each number in the center of the card and one inch high, using your best "first grade" handwriting. [In other words, write plainly and without any slant.]

Source for Free 100 Chart

https://allfreeprintable.com/blank-100-chart

There are many online sources for printable 100 charts. Here's one. Because I want the students to add their own colors, I like this one because it is in black and white and very plain. By the end of First Grade, some children are counting past 100. So, toward the end of the year, you might consider printing out a blank chart that goes up to 120, or even two blank 100 charts so your child can write the numbers up to 200.

Pencil Grip and Posture

The ability to pay attention, participate in class, and learn material depends to an unexpected degree on sitting and holding the pencil properly. For this reason, I emphasized how to set up your classroom in the introductory materials.

Audrey McAllen, who basically founded the Waldorf "Extra Lesson" program, was quite clear on the importance of good posture. She described a child's desk and chair as follows:

- The child needs to be able to sit in the chair with both feet flat on the floor, knees a little higher than the hips and the spine curved so that "a crown could balance on the head."
- The height of the desk should allow the writing arm to rest comfortably on it without pushing up the shoulder. The non-writing hand should rest on the side of the paper, keeping it in place and moving it so that the arm remains in the correct position.

The pencil itself should be held about one inch above the point, between the thumb and forefinger of the writing hand. Only a mild pressure is necessary to hold the pencil in place—if you see white knuckles, your child is holding the pencil too tightly. The pencil should rest just below the first knuckle of the middle finger; all knuckles should be gently rounded. The elbow should be stationary.

The ability to hold a pencil correctly proceeds in stages, but by the time a child is ready for First Grade, they should be easily holding the pencil as described here. However, many children today struggle with holding a pencil correctly and sitting properly. Some researchers have correlated these weaknesses with an inability to pay attention and/or to control emotions. If your child finds either skill challenging, you might consider consulting an occupational therapist for helpful exercises.

Poems and Songs

These first songs and poems will start every lesson for the entire year.

Through the Night [song]

Through the night, the Angels kept
Watch beside me while I slept.
Now the dark has gone away.
We are glad for this new day.

Morning Has Come [song]

Morning has come, night is away.
We rise with the sun and welcome the day.

Round and Round [song with movement]

Round and round, the Earth is turning,	*[Standing straight and tall, turn slowly in*
Turning always round to morning,	*place until you are facing forward at the*
And from darkness round to light.	*end of the song.]*

Tall Trees in the Forest [verse with movement]

[Repeat three times, varying speed of movement once the child knows the verse.]

Tall trees in the forest;	*[Standing straight and tall, lift up on toes and reach high with your arms.]*
Pine cones on the ground.	*[Slowly bend over and "touch" the pine cones.]*
Tall trees in the forest,	*[Standing straight and tall.]*
They bend, they bend, they bend.	*[Bend over slowly and touch the floor.]*

[Various songs and poems will be inserted here for each series of 6 lessons.]

These verses will end the Morning Activities and be a signal that the seated work of Morning Lesson is about to begin.

Verse with no movement:
[Stand up straight and tall, crossing arms over the heart, with reverence.]

The Sun with loving light makes bright for me each day.
The Soul with Spirit power gives strength unto my limbs.
In sunlight shining clear, I do revere, O God,
The strength of humankind which you so graciously
Have planted in my soul,
That I with all my might
May love to work and learn.
From you stream light and strength.
To you rise love and thanks.

I Look at My Hands [verse with gestures]

I look at my hands with my fingers fine,	*[Hold hands out in front of you.]*
And I want to be proud that they are mine.	
For deep in my heart lies a golden chest	*[Cross hands over heart.]*
With secret treasures that no one can guess	
Unless my hands do their very best.	*[Hold hands out in front of you.]*

Warm Our Hearts [song to end Morning Lesson]

Warm our hearts, O Sun, and give	*[Cross hands over heart.]*
Light that we may daily live,	
Growing as we want to be:	
True and good and strong and free.	

These next poems and songs will change periodically. The first ones listed are the ones we recommend for the beginning of the year. After the first few lessons, some will be put aside and possibly brought back later in the year. New songs and poems will be added.

Use these poems and songs for the first six classes:

Humpty Dumpty [verse with movement]
[Walk slowly forward, stepping on the accented syllables. Then reverse and walk backwards.]

Humpty **Dump**ty **sat** on a **wall**.
Humpty **Dump**ty **had** a great **fall**.
All the King's **hors**es, and **all** the King's **men**
Could **not** put **Hump**ty to**geth**er a**gain**.

Straight as a Spear [verse with movement]

Straight as a spear, I stand.	*[Standing as tall and straight as possible.]*
Strength fills my legs and arms.	*[Spread legs apart, stretch arms straight out to sides.]*
Warmth fills my heart with love.	*[Cross arms over heart.]*

[Swinging your arms around to each side, jump back into upright position, arms at sides.]

Stepping Stones [verse with movement]
[Step forward on accented syllables. Notice that some steps are slow, some faster.]

Step on **step**ping stones, **one, two, three**.
Step on **step**ping stones, **fol-low me**.
The **river** is **fast,** the **river** is **wide**.
We'll **step** on **step**ping stones **to** the **other side**.

I Am a Tall Tree [verse with movement]

I am a tall tree, reaching up to the sky.	*[Standing straight, stretch arms up.]*
When the wind blows, I lean with a sigh.	
I lean to the front,	*[Legs straight, lean to the front.]*
I lean to the back,	*[Legs straight, lean to the back.]*
I stand up straight without a crack.	*[Jump back into uprightness.]*

Pease Porridge Hot [verse with finger movement]

[Touch thumb to fingertips in turn on accented syllables, first with right hand, then with left, then both together. Start with thumb to index finger; continue to little finger, then go

back one fingertip at a time to the index finger. Repeat. At the end of the verse, you will have touched each fingertip with your thumb twice.]]

Pease por**ridge hot**, pease por**ridge cold**,
Pease por**ridge in** the **pot, nine days old.**

For the second set of 6 classes, omit "Humpty Dumpty." Use these songs and poems:

The Hokey-Pokey [dance song with gestures, as indicated]

You put your hand in, you take your hand out,
You put your hand in and you shake it all about.
Do the hokey-pokey and you turn yourself around:
That's what it's all about! *[Clap and stomp in rhythm on each syllable .]*

You put your other hand in.....

You put your foot in......

You put your other foot in......

The "dance" can continue for another few verses, but be careful not to spend the entire lesson on the Hokey-Pokey! Other body parts to put in: knee, elbow, back, head. I did not use "left" and "right" because first grade children might be able to identify these at the beginning of the year. However, feel free to use them if your child will not be confused.

Straight as a Spear [as above]

Stepping Stones [as above]

Double, Double   *[Try this at varying speeds.]*

Double, double	*[Form hands into fists & thump one on top of the other; switch.]*
This, this.	*[Push the back of each hand forward, in rhythm.]*
Double, double	*[Repeat as above.]*
That, that.	*[Push the front of each hand forward, in rhythm.]*
Double this.	*[One fist bump, then the backs of the hands forward.]*
Double that.	*[The opposite fist bump, then the fronts of the hands forward.]*
Double, double, this, that.	*[Fist bumps, back of hands, front of hands.]*

Count to 10, bouncing a play ball on each number and holding it in your hands on the pause between numbers.

I Am a Tall Tree [as above]

For the first 6 lessons of Block B [Quality of Number], omit "The Hokey-Pokey" and add these songs and poems.

This Old Man [Traditional Children's Song]
[I usually play this game sitting on the floor, but you can play it standing or sitting in a chair.]

This old man, he played one.	*[Hold up one finger.]*
He played knick-knack on my thumb. [	*Point to thumb.]*

Chorus:

With a knick-knack	*[Clap on Knick, pat knees on Knack.]*
Paddy-whack	*["]*
Give a dog a bone	*[Pretend to throw a bone.]*
This old man came rolling home.	*[Roll hands over lap while swaying side to side.]*

For the remaining verses, use the same gestures. Sing the chorus after each verse.

This old man, he played two.
He played knick-knack on my shoe.

This old man, he played three.
He played knick-knack on my knee.

This old man, he played four.	
He played knick-knack on my door.	*[Hold one hand up, palm flat, as the "door."]*

This old man, he played five.	
He played knick-knack on a beehive.	*[Hold up one fist as the "beehive."]*

This old man, he played six.	
He played knick-knack on some sticks.	*[Splay fingers and cross one hand over the Other.]*
This old man, he played seven.	
He played knick-knack up to Heaven.	*[Point upwards.]*

This old man, he played eight.	
He played knick-knack on my gate.	*[Hold up one hand, palm flat, & swing back and forth.]*

This old man, he played nine.
He played knick-knack up a vine. *[Pretend to climb a vine.]*

This old man, he played ten.
He played knick-knack all over again. *[With hands and face, show dismay!]*

You Are One [by Vivian Jones-Schmidt

You are **one** and **I** am **one**, *[Bounce ball on accented syllables.]*
And **one** is the **gol**den **Sun**.

Golden **Sun** and **Moon** make **two**,
And **so** do **I** and **you**.

Sun and **Moon** and **Earth** will **be**
Always to**geth**er and **always three**.

Fall and **Winter**, **add** two **more**:
Spring and **Sum**mer make **seasons four**.

I am a **star** that **shines** so **bright**,
With **five** strong **points** that **make** my **light**.

Count to ten and back to one, bouncing the playball on each bounce and holding it in your hands in the pause between numbers.

I Am a Tall Tree, as above.

For the second set of lessons in Block B, insert these activities.

One, Two, Buckle My Shoe *[with a play ball]*
[Bounce the ball on each accented syllable, and catch on the unaccented syllable.]

One, **two,** buckle my **shoe.**
Three, **four,** shut the **door.**
Five, **six,** pick up **sticks.**
Seven, **eight,** lay them **straight.**
Nine, **ten,** a big fat **hen.**
Eleven, **twelve,** dig and **delve.**
Thirteen, **fourteen,** dishes need **sorting.**

Fifteen, **sixteen,** take them to the **kitchen.**
Seventeen, **eighteen,** we're all **waiting.**
Nineteen, **twenty,** there's work a-**plenty.**

Count to twenty, walking forward one step on each number. Now count back from twenty to one, taking one step backward on each number. Steps can be very small and very loud!

Bean bag Exercises
[There are seven bean bag exercises in the sequence, but we will start with one. Each exercise is accompanied by a poem.]

Cross-patch
Cross patch, **draw** the **latch.**   *[Hold the bean bag in your right hand, chin-height above the left hand, which is under the right hand and at waist level. Drop the bean bag into the left hand on the accented syllable. Now switch places with the hands. Drop the bean bag from the left hand [on top] into the right hand [underneath] on the accented syllable. Switch hands again and continue to drop the bean bag into the lower hand on the accented syllable.]*
Sit by the fire and **spin.**
Take a cup and **fill** it up.
In**vite** the neighbors **in.**

Insert these activities in the Morning Activities for the first six lessons of Block C.

Clap Hands *[traditional children's song]*
Clap, clap, clap your hands,   *[Do the motions as you sing the words.]*
Clap your hands together.
Clap, clap, clap your hands,
Clap your hands together.

Stomp your feet.....

Nod your head.....

Jump up high.....

Turn around.....

[Add any motions you or the child think of!]

Bean Bags

Cross-patch, draw the latch,   *[motions as before]*
Sit by the fire and spin.
Take a cup and fill it up,
Invite the neighbors in.

Whisky, frisky, hippety-hop!   *[Start with the bean bag held in front and toss it*
Up he goes, to the tree top!   *around your body in time with the verse.]*
Whirly, twirly, round and round,
Down he scampers to the ground.
Furly, curly, what a tail!
Tall as a feather, broad as a sail.
Where's his supper? In the shell.
Snappy-cracky, out it fell.

One Shoe, Two Shoes

One shoe, two shoes   *[1. Tap left foot with right hand in front, then right foot with left hand, also in front.]*

Three shoes, four.   *[2. Tap left foot with right hand, behind; then right foot with left hand, also behind.]*

Check our pony's silver shoes,   *[Pattern #1.]*
Let's tap them all once more.   *[Pattern #2.]*

One shoe, two shoes,   *[Repeat as above.]*
Three shoes, four.
Let's go riding, pony dear,
Through the stable door.

Clippety-Clop

[Pretend you are horses, trotting through the countryside. Follow the directions.

Clip-clop, clippety-clop,
To the countryside, don't stop!
Clip-clop, clippety-clop,
Faster, faster—slower, slower–
Whew! It's time to stop!

Counting with Play Ball

Hold the ball in your hands on the odd numbers;
Bounce the ball on the even numbers and speak louder.
Count to 12; stop. Count back from 12 to 1.

Do this several times, varying the speed and also the volume of your voice. For example, try whispering on the odd numbers and shouting on the even numbers.

When your child is comfortable with the numbers 1 - 12, keep going to 24, forward and back.

Insert these activities into the Morning Lesson for Lessons 7 - 12 in Block C.

Stomping Land [traditional children's song]

I traveled far across the sea,	*[Put hand to forehead, look into distance]*
I met a man, and old was he.	*[One hand on "cane," the other on bent back]*
"Old man," I said, "Where do you live?"	*[Hands on hips.]*
And this is what he told me:	*[Index finger 'bouncing' for emphasis]*
"Follow me to stomping land,	*[Stomp]*
Stomping land, stomping land.	
All who wish to live with me,	
Follow me to stomping land.	*[In successive verses, choose an action and do it.]*

Bean Bags

Cross-patch, draw the latch,	*[motions as before]*
Sit by the fire and spin.	
Take a cup and fill it up,	
Invite the neighbors in.	
Whisky, frisky, hippety-hop!	*[Start with the bean bag held in front and toss it around your body in time with the verse.]*

Up he goes, to the tree top!
Whirly, twirly, round and round,
Down he scampers to the ground.
Furly, curly, what a tail!
Tall as a feather, broad as a sail.
Where's his supper? In the shell.
Snappy-cracky, out it fell.

Lovely *rain*bow, **hung** so *high*, switch
Glistening *in* the **clear**ing *sky*.

Red and *orange*, **yel**low and *green*,
Rainbow *colors* **bright**ly *gleam*.

Blue and *indigo*, **pur**ple so *fine*,
Rainbow *colors* **soft**ly *shine*.
Pots of *gold* may **there** be *found*,
Where the *colors* **meet** the *ground*.

*[Bring hands together over head and bean bag from one hand to the other on **bold** syllables.*

On italic syllables, hold hands out to sides, arms parallel to floor.]

Teddy Bear *[Follow instructions in verse for motions.]*
Teddy Bear, Teddy Bear, turn around.
Teddy Bear, Teddy Bear, touch the ground.
Teddy bear, teddy bear, reach up high,
Teddy bear, teddy bear, touch the sky,
Teddy bear, teddy bear, bend down low,
Teddy bear, teddy bear, touch your toes,
Teddy bear, teddy bear, go to bed,
Teddy bear, teddy bear, rest your head,
Teddy bear, teddy bear, turn out the lights,
Teddy bear, teddy bear, say "good night".

You Are One [by Vivian Jones-Schmidt]
*[Hold up the correct number of fingers for each verse.
For 11 and 12, hold up ten and then one or two more, as needed.]*

You are one and I am one,
And one is the Golden Sun.

Golden Sun and Moon make two,
And so do I and you.

Sun and Moon and Earth will be
Always together and always three.

Fall and Winter, add two more.
Spring and Summer make seasons four.

I am a star that shines so bright,
With five strong points to make my light.

The cells within each hive of bees
With six sides hold the honey sweet.

Within each rainbow in the sky
Seven colors stretch side to side.

Sparkling in the morning dew
Spider's eight legs weave a web that's new.

Nine are the candles glowing bright
That sit in the window in the dark night.

Ten fingers have I, five on each hand,
And ten strong toes that help me stand.

Eleven hours of sleep at night
Keep me happy in the sunshine bright.

Every year is twelve months long,
And each month sings a different song.

Counting with Play Ball

Count to 24, emphasizing the even numbers.
Hold the ball in your hands on the odd numbers, and bounce on the even numbers.
When your child can do this without hesitation, whisper on the odd numbers and shout on the even numbers.
Count forward from one to twelve and then back from 12 to 1.

Insert these activities into the Morning Exercises for Lessons 1-6 of Block D.

Bean Bags

#1. Cross-patch, draw the latch,   *[motions as before]*
Sit by the fire and spin.
Take a cup and fill it up,
Invite the neighbors in.

#2. Whisky, frisky, hippety-hop!
Up he goes, to the tree top!
Whirly, twirly, round and round,
Down he scampers to the ground.
Furly, curly, what a tail!
Tall as a feather, broad as a sail.
Where's his supper? In the shell.
Snappy-cracky, out it fell.

[Start with the bean bag held in front and toss it around your body in time with the verse.]

#3. Lovely *rain*bow, **hung** *so high,*
Glistening *in* the **clear**ing *sky.*
Red and *orange,* **yel**low and *green,*
Rainbow *colors* **bright**ly *gleam.*

Blue and *in*digo, **pur**ple so *fine,*
Rainbow *colors* **soft**ly *shine.*
Pots of *gold* may **there** be *found,*
Where the *colors* **meet** the *ground.*

*[Bring hands together over head and switch bean bag from one hand to the other on **bold** syllables. On italic syllables, hold hands out to sides, arms parallel to floor.]*

#4. Water, **wa**ter, **wa**ter **fall.**
Wher**ev**er I **am,** I **hear** your **call.**
You are **love**ly, **spark**ling, **tall.**
You **make** of **wa**ter, **one** strong **wall.**
#5. Jack be **nim**ble, **Jack** be **quick!**
Jack jump **ov**er the **can**dle **stick!**

[Hold bean bag in right hand behind head. On bold syllables, drop bean bag into your left hand, which is held at mid-back, cupped to catch the bean bag.]
*[Hold bean bag in the right hand. On the bold syllables, switch hands **under** your left knee. Then, with the bean bag in the left hand, switch to the right hand under the right knee. Continue the pattern, varying the Speed.]*

Copper Rod Exercises:
These introductory copper rod exercises "wake up" the hand and arm muscles needed for writing and other skills.

#1: Hold the rod in both hands, palms up. Reciting "Wee Willie Winkie," gently roll the rod back and forth, catching it with the thumb so that it does not progress beyond the hand.

Wee Willie Winkie runs through the town,
Upstairs and downstairs in his nightgown.
Rapping at the windows, crying through the locks:
"Are the children all in bed? For now it's eight o'clock!"

#2: This time, recite "Hickory Dickory Dock" as you allow the rod to gently roll all the way to your elbows and back. [Feel free to make up your own verses, as I have here!]
Hickory Dickory Dock, the mouse ran up the clock.
The clock struck one, the mouse ran down,
Hickory Dickory Dock.
Hickory Dickory Dock, the mouse ran up the clock.
The clock struck two, the mouse sneezed, "Achoo!"
Hickory Dickory Dock.
Hickory Dickory Dock, the mouse ran up the clock.
The clock struck three, the mouse called, "Whee!"
Hickory Dickory Dock.

#3. Now sing "Row, Row, Row Your Boat" while allowing the rod to roll all the way up to your shoulders and back to your fingertips.
Row, row, row your boat gently down the stream.
Merrily, merrily, merrily, merrily: Life is but a dream.

#4. Humpty Dumpty
[Start with the copper rod held horizontally at waist height in both hands, palms down. With the first two lines, raise the rod over your head; with the second two lines, lower the rod as far as you can without bending over.]

Humpty **Dump**ty
Sat on a **wall.**
Humpty **Dump**ty
Had a great **fall.**
[Start these next two lines holding the rod horizontally at the waist. Without shifting your hands from the previous position, first move the rod to the vertical with the right hand on top, then the left.]
All the King's **horses**
And **all** the King's **men**
[The rod is once again at waist height. On the first line, lift it over your head; on the second line, lower it as before.]

Could **not** put **Hump**ty
To**get**her a**gain**.

#5. Now place the rod carefully on top of your head and, walking slowly, recite this verse while following the directions for additional movement. This is a challenge!

The Stork

I lift my leg, I stretch my leg,
I plant it, firm and light.
I lift again, and stretch again,
My pace, exactly right.
With care I go, so grand and slow–
I move just like a stork.
My eye is bright, my head upright,
And pride is in my walk.

#6. Hold the copper rod with both hands, palms down, in front of you. Reciting this poem, lift one finger at a time as indicated. Start with your right hand, then lift the fingers of the left hand, then move the fingers of both hands simultaneously. As you become accustomed to the pattern, vary the speed.

Peter	*[index finger]*
Peter	*[middle finger]*
Pumpkin	*[ring finger]*
Eater	*[little finger]*
Had a	*[little finger]*
Wife but	*[ring finger]*
Couldn't	*[middle finger]*
Keep her.	*[index finger]*
He put her	*[index finger]*
In a	*[middle finger]*
Pumpkin	*[ring finger]*
Shell	*[little finger]*
And there	*[little finger]*
He kept her	*[ring finger]*
Very	*[middle finger]*
Well.	*[index finger]*

Insert these activities into the Morning Exercises for Lessons 7 - 12 of Block D.

Jump Rope

Ideally, by now your child can jump rope independently–and you can, too! Here are some counting rhymes for jumping rope. You might also know of others. Choose two or three and alternate them by the week. For each rhyme, you can choose to practice counting by ones, twos, threes, etc.

Over in the meadow where the green grass grows,
There sat [child's name] as sweet as a rose.
Along came a frog and kissed her/him on the nose.
How many kisses did s/he get?

Cinderella, dressed in yellow
Went downstairs to kiss her fellow.
How many kisses did he get?

1-2-3-4-5-6-7
All good children go to Heaven.
7-6-5-4-3-2-1
Jumping rope sure is fun!

Bubble gum, bubblegum in the dish,
How many pieces do you wish?   [traditional]

Hickety pickety pop!
How many times before I stop?   [Ellen Mason]

Wash the dishes, dry the dishes,
Have a cup of tea.
Don't forget the sugar–
1, 2, 3, etc.   [Erin Pollard]

Play Ball

Choose a multiplication table or addition set to practice, rhythmically, with your ball.

Examples:

Three	*[bounce ball]*
Is	*[catch ball]*
Three times	*[bounce ball]*
One.	*[catch ball]*

[Continue with: 6 is three times two, nine is three times three, etc., up to 36.]

Two	*[bounce ball]*
Is	*[catch ball]*
One plus	*[bounce ball]*
One.	*[catch ball]*
Three	*[bounce ball]*
Is	*[catch ball]*
One plus	*[bounce ball]*
Two.	*[catch ball]*

[Continue with: 4 is one plus three, 5 is one plus four, etc. You can also reverse the numbers, as in: 4 is three plus one, 5 is four plus one, etc. Play with it.]

Count backwards from 12, bouncing your ball on each number and pausing as you catch it.

Copper Rod Exercises

#4. Humpty Dumpty

[Start with the copper rod held horizontally at waist height in both hands, palms down. With the first two lines, raise the rod over your head; with the second two lines, lower the rod as far as you can without bending over.]

Humpty **Dump**ty
Sat on a **wall.**
Humpty **Dump**ty
Had a great **fall.**

[Start these next two lines holding the rod horizontally at the waist. Without shifting your hands from the previous position, first move the rod to the vertical with the right hand on top, then the left.]

All the King's **horses**
And **all** the King's **men**
[The rod is once again at waist height. On the first line, lift it over your head; on the second line, lower it as before.]
Could **not** put **Hump**ty
To**geth**er a**gain.**

#5. Now place the rod carefully on top of your head and, walking slowly, recite this verse while following the directions for additional movement. This is a challenge!

The Stork

I lift my leg, I stretch my leg,
I plant it, firm and light.
I lift again, and stretch again,
My pace, exactly right.
With care I go, so grand and slow–
I move just like a stork.
My eye is bright, my head upright,
And pride is in my walk.

#6. Hold the copper rod with both hands, palms down, in front of you. Reciting this poem, lift one finger at a time as indicated. Start with your right hand, then lift the fingers of the left hand, then move the fingers of both hands simultaneously. As you become accustomed to the pattern, vary the speed.

Peter	*[index finger]*
Peter	*[middle finger]*
Pumpkin	*[ring finger]*
Eater	*[little finger]*
Had a	*[little finger]*
Wife but	*[ring finger]*
Couldn't	*[middle finger]*
Keep her.	*[index finger]*
He put her	*[index finger]*
In a	*[middle finger]*
Pumpkin	*[ring finger]*
Shell	*[little finger]*
And there	*[little finger]*
He kept her	*[ring finger]*
Very	*[middle finger]*
Well.	*[index finger]*

Samples of Title Pages

Title Page Instructions

Shown below are samples of title pages I've used. The first is very simple, displaying only the words; while the rest range from an illustration to decorative motifs. The adult will need to plan the overall layout of the page before bringing the project to the child. You can use the first page in the book for your plan if you work in light graphite pencil.

- I used the same sequence for each title page:
- Using block crayons, measure across the top of the page to the center.
- Mark that spot, and draw a vertical line down the center.
- As you have already planned the words you'll write, divide those words into the number of lines you want to use. Determine where the lines will be by measuring with a block crayon and marking with your graphite pencil.
- Determine the necessary length of each line by writing out the words [in your best "teacher handwriting"] on each line.
- Guide the child through this process.

I recommend writing the words and drawing the "helping lines" with a colored pencil. For example, the "helping lines" on the finished page could be in yellow or gold, while the words could be in a darker color.

The specific words you use will be ones that you choose. Here are a couple of examples:

ARITHMETIC LESSONS	ARITHMETIC PRACTICE
FIRST GRADE	AND FORM DRAWING
DATES	DATES
CHILD'S NAME	CHILD'S NAME

When you've finished writing the words, create the background. You can choose any background you like. I recommend that you give this some thought so the finished page will be pleasing to the eye.

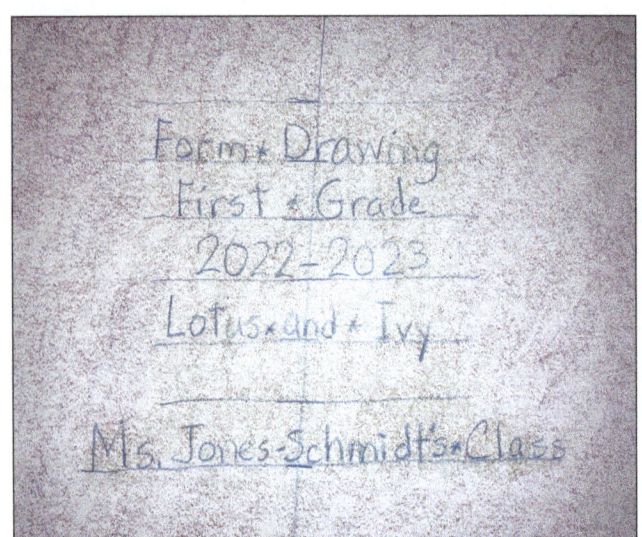

Supplemental Activities

Counting
- How many apples, bananas, berries, oranges are in your refrigerator?
- How many chairs are in your home?
- How many windows are in your home?
- How many doors are in your home? How many are inside doors?
- How many steps are in your home?
- How many steps do you make to get to the park? The fence? A store? A neighbor's house?
- How many people on the bus are wearing red?
- How many trees do you see out a window?

Division
- We have 15 berries and three people. How many berries will I give to each person?
- We have 6 books and 2 children. How many books can each child read?
- We have 9 pieces of paper and 3 children. How many pieces of paper will I give to each child?

Bean Bag
Of course you can purchase bean bags for yourself and your child, just making sure they are no larger than 2.5." This is the most comfortable size for your child to hold without dropping. Bean bags are also simple to make.

Materials for a 2" beanbag
- Paper and pencil
- Scissors
- Fabric
- pins
- Thread
- Needle
- Dry Lentils

- Draw a square 2.5" on each side. Lay this paper pattern on your fabric and cut out two squares.
- Pin them together with the wrong side of the fabric facing out.
- Thread your needle; sew around the sides, leaving a hole about 1" wide on the fourth side. Your bean bag will be much sturdier if you stitch around the sides twice.
- Turn your bean bag right side out.
- Fill it loosely with the dry lentils.
- Stitch the opening closed.

Supplemental Sources for Activities

There are many online sources for arithmetic materials and ideas. Here are a few that I've found to be helpful. However, as it's important not to confuse your child, I recommend that you wait until you have finished this course as written before exploring other resources. That way, you can choose activities that match what you've taught.
For example, the traditional way to teach addition is: 4 + 3 = ____. We teach: 7 = 4 + ____, starting with the whole. The traditional way to teach subtraction is: 9 − 5 = ____. We teach: 9 − ___ = 4.
We also avoid worksheets in First Grade, and many websites focus on worksheets.

https://buildmathminds.com/ This is a good source for dot cards, etc.
https://www.teacherspayteachers.com/Product/FREE-Subitizing-Dot-Cards-2042252
https://www.youcubed.org/tasks/ This is a good source for games.
https://www.weareteachers.com/math-card-games/
https://familymath.stanford.edu/activities/card-games/
https://www.education.com/

https://www.myteachingstation.com/worksheets/kindergarten/writing-numbers [This is a good source for printable templates for practicing writing the numbers 0 - 9.

String Games are fantastic for building eye-hand coordination and cognitive skills. Here is one book I've found helpful:
Cat's Cradle, Owl's Eyes: A Book of String Games, by Camilla Gryski
There are also instructional videos on YouTube.

The Temperaments

Rudolf Steiner believed and taught that every human being has the capacity to embark on a path of self-development. To help in this journey, Steiner developed several frameworks. The four temperaments, which he traced to the ancient Greeks, were one of these. Although Steiner used the Greek nomenclature for the temperaments, his descriptions and suggestions on how to use them were very different.

There are four basic temperaments: phlegmatic, melancholic, sanguine, and choleric. Ancient Greeks used these descriptors in medicine to diagnose and treat illness. In Steiner's teaching, understanding one's temperament can be a first step in understanding one's motivations and actions. This can also be helpful in understanding another person. Thinking about another person's temperament is not a judgment of that person; nor is it a conclusion you might reach and announce to that person. We use the schema of the temperaments in an effort to understand the children we teach [and perhaps the adults we work with], but this is a tool that is not explained to the children.

First, I'll indicate how to imagine the temperaments. Each one can, for example, be seen as a color, an animal, one of four "elements," a season, and a tree. That's how I'll start in my presentation of each temperament, beginning with the melancholic.

Although *"melancholic"* might seem to indicate that persons of this temperament are always sad, this would not be accurate. The melancholic temperament is related to the EARTH in terms of the elements, because the earth is generally stable and unmovable. Melancholics have a hard time making decisions, for example, because they keep looking at every option and weighing the pros and cons. They can thus be perceived as being inflexible, as change is difficult for them. The color associated with this temperament is blue, and the season is Autumn. The associated tree is of course the weeping willow, and the animal a hound. Melancholics often drag their feet when walking, heads down and deep in self-absorbed thought. They can be very sensitive and easily hurt, and they love sympathy. A melancholic never forgets an injury or insult and thinks that "bad things" only happen to them. If you know the book "Winnie the Pooh," the character of Eeyore possesses a perfect melancholic temperament.

The *"phlegmatic"* temperament is related to the element of WATER. Unlike the earth or stones, water moves; but it is hypnotic in its movement, and a phlegmatic person often seems to have been hypnotized into another, more pleasant world. The associated color is green, and the season is Winter. The associated animal is the cow, and the tree is the sequoia, which grows very slowly but with great steadiness. A person of phlegmatic temperament walks dreamily and is generally very placid. They like routine and personal comfort, needing time to adjust to any change. Hence they can be perceived as stubborn. Phlegmatic persons are methodical and slow in their approach to any task. In terms of "Winnie the Pooh," think of Pooh, himself.

The *"sanguine"* temperament is related to the element of AIR. Consider the wind, the "merry little breezes" of Old Mother West Wind, and you'll have the general idea. Persons of sanguine temperament have a hard time focusing on any one thing for very long, because everything is so interesting! The season associated with the sanguine is Spring, and the color is yellow. Sanguine persons are sociable and chatty, if a little superficial. They move by skipping, or by appearing to skip, and sometimes hardly seem to touch the ground. Sanguines are incurable optimists who cannot stay angry or sad for very long. The associated animal is the squirrel, and the tree is the birch, always open to the vagaries of the wind. In terms of "Winnie the Pooh," remember the character of Tigger.

The *"choleric"* temperament is associated with the element of FIRE, and persons in whom this temperament dominates are energetic and active, focused on directing the action. They walk with firm steps and can be perceived as being aggressive. The season associated with the choleric is summer, and the color is red. A choleric person can "flame" into rage, always ready to blame any failure on someone else. The associated animal is, not surprisingly, the bull, and the tree is the oak. In the story of "Winnie the Pooh," the character is that of the Rabbit.

Considering all four temperaments, there are two which characterize introverts [the phlegmatic and the melancholic] and two which characterize extroverts [the sanguine and the choleric]. Steiner characterized each temperament in terms of its relation to memory:

- The melancholic observes little, but remembers it especially if it concerns them.
- The phlegmatic learns slowly, but remembers everything if sufficiently awake.
- The sanguine notices everything, but remembers nothing.
- The choleric observes what is of interest to them, and then forgets it.

These characterizations can be very helpful to the teacher, who then must figure out how to best work with children who demonstrate each temperament.

Thankfully, we can look to those who have gone before us! Again, for example, Roy Wilkinson has solid advice for working with the temperaments in children. With the melancholic, who can be very attuned to any kind of suffering, it might help to explain how others will suffer if the child does not do as asked. Melancholics also respond to a sympathetic adult. Phlegmatics must be shocked into paying attention, so the teacher can speak very clearly and succinctly [and perhaps loudly!] while being calm and strong.

To encourage the compliance of the sanguine, Wilkinson advises us to "ask a personal favor," while being firm, friendly, and interested in the child's activity. To stimulate the choleric, present a challenge; cholerics respond to strength and "plain talk."

There's an old story which goes far in distinguishing among the temperaments. Imagine there's a large stone in a path. As a melancholic approaches the stone, they begin to wail, "Why is this stone in MY way? It's terrible! What can be done?"Continues to weep. The phlegmatic approaches the stone and unconsciously thinks, "What a nice rock. I'll sit down." Sits on the stone and dreams away the day. The sanguine approaches the stone and doesn't even notice it, but skips right over it. The choleric approaches the stone and kicks it out of the way: "How dare you sit in my path?"

Steiner was very clear in stating that no one is a "pure" temperament: each person is a blend of two or more. As adults on paths of self development, we might start with self-reflection: "What is my dominant temperament? What other temperaments are active in my personality?" And, as we will each ideally develop all four temperaments, we can ask ourselves," Are there temperaments I need to develop?" Each temperament offers positive and helpful qualities. The choleric is a strong leader, energetically marching forward; the sanguine has a sunny disposition and is interested in everything; the phlegmatic is a deep thinker, calm and reflective; and the melancholic is sensitive to anyone's pain. If we have access to the characteristics of all four temperaments, we are free to consider each situation and choose which temperament to call upon.

So, how are the temperaments related to arithmetic? In the first training for teachers, Steiner noted that "...I find it possible to make use of the four rules of Arithmetic to arouse the interest of the four temperaments. Adding is related to the phlegmatic temperament, subtracting to the melancholic, multiplying to the sanguine, and dividing, working back to the dividend, to the choleric." *[Discussions with Teachers, lecture 4]* Steiner was quite clear that "we must always take care that, as teachers, we create what goes from us to the children in an exciting way so that it gives rise to imagination." *[The Foundations of Human Experience]* I believe that the combination of these two elements–temperaments and imagination–inspired the creation of the "gnome stories" so prevalent in the early grades of many Waldorf schools. You might know of these stories, as each operation of arithmetic is given a particular color, which is worn by a little gnome which enacts stories utilizing that particular operation. Thus, for example, addition is represented by a gnome dressed in green and subtraction by a gnome dressed in blue. The stories can be enchanting; but it is easy to become so involved with the gnomes that one loses track of the operation itself. For this reason, I have advocated working with children dressed in the various colors, rather than with gnomes. Gnomes, and the work they do in the earth, are perhaps more appropriate for adult contemplation.

Summary of Temperament Attributes

from the work of Hippocrates, Galen, Rudolf Steiner, and Roy Wilkinson

CHOLERIC

fire	**dry/hot**	**summer**	**red**
RABBIT [Milne]	wants to lead	aggressive	firm step

- active, energetic
- can rage and go into tempers — kicks stone out of way
- any failure is someone else's fault — bull — oak
- observes what is of interest to them and then forgets it
- more intensely extroverted

SANGUINE

air	**hot/moist**	**spring**	**yellow**
TIGGER [Milne]	sociable	chatty/superficial	skips

- dominated by sensations, ideas, feelings — over stone
- cannot fix attention or hold onto an idea — squirrel — birch
- incurable optimist, cannot stay angry for long
- notices everything but remembers nothing
- extroverte

INTROVERTED		EXTROVERTED	
Melancholic	*Phlegmatic*	*Sanguine*	*Choleric*
more intense	less intense		more intense

- Melancholic: fall, blue, earth
- Phlegmatic: winter, green, water
- Sanguine: spring, yellow, air
- Choleric: summer, red, fire

PHLEGMATIC

water	cold/moist	winter	green
POOH [Milne]	methodical	placid	dreamy walk

- desire to maintain inner comfort — sits on stone without noticing it
- likes routine, cannot leave any task unfinished but slow
- needs time to adjust to any change — cow — sequoia
- can be stubborn
- learns slowly but remembers everything if sufficiently awake
- honest, reliable, conscientious
- introverted

MELANCHOLIC

earth	dry/cold	fall	blue
Eeyore [Milne]	inflexible	easily hurt	dragging walk

- think challenging experiences only happen to them
- never forgets an injury or insult — hound — weeping willow
- wants sympathy Why is this stone in MY way? Terrible! What's to be done?
- observe little but remember it, especially if it concerns them
- drooping head, downward glance, sad eye, self-absorbed
- can be very sympathetic and helpful if sympathy is engaged
- more intensely introverted

The Temperaments Outline

History—for context

- founding of first Waldorf School/ Emil Molt
- Hippocrates
- Galen
- VISUAL: temperaments with elements, qualities
- development of temperaments through life

PHLEGMATIC

- Pooh: pp. 129-131
- Characteristics

CHOLERIC

- Rabbit: pp. 39-40
- Characteristics

SANGUINE

- Tigger: pp. 22-25
- Characteristics

MELANCHOLIC

- Eeyore: pp. 11-13
- Characteristics

SITUATIONS

PHLEGMATIC

- *If you lose something, just keep looking.*
- *Stone in your way? Just sit on it.*

SANGUINE:

- *What were you looking for again?*
- *What stone? [skips past]*

CHOLERIC:

- *You'd better find it.*
- *Kick the stone out of the way.*

MELANCHOLIC:

- *You'll never find it.*
- *Someone put this stone here just to be in my way. Sits on stone and mourns.*

CHOLERIC
fire

[dryness]   *[heat]*

MELANCHOLIC   **SANGUINE**
earth   *air*

[cold]   *[moisture]*

PHLEGMATIC
water

Steiner on Arithmetic

The Kingdom of Childhood

Page 90: Counting should be derived from life itself....

— the unit of one person: you can chop a piece of wood into two pieces, but you cannot do that with a person.
— you have two hands; they can touch one another; they can come together; this is different from when you move alone—you move as a unit; this is a duality: "two"
— two children can walk toward one another; if another joins them, they are three; you cannot make three with your hands only
— you can find two somewhere else in your body; legs and feet: does a dog have only two feet? No, it has four.

Number must be built up out of life.
Start with Roman numerals: I, II, III, IIII, V—using the fingers and thumb.

Only when number has been worked out from life should you try to introduce counting by letting the numbers follow each other. The children should be active here. Start with a rhythm: 1,2; 1,2; 1,2, with a stomp on 2. Continue with 1, 2, 3.

Kingdom of Childhood
page 93:

The head is a metamorphosis of the former life on earth, and the fact of having a head only begins to have a real meaning for man when he knows something of his former earth lives. All other activities come from somewhere else, not from the head at all. The truth is that we count subconsciously on our fingers...And what you do in this manner with your fingers and toes only throws its reflection into the head. The head only looks on at all that occurs. The head in man is really only an apparatus for reflecting what the body does. The body thinks, the body counts. The head is only a spectator.....All that is done in spiritual life is done from the body. Mathematics are done by the body, thinking is also done by the body, and feeling too is done with the body.

Thus from the most varied sides you must try to build up what the child has to learn as counting. And when you have worked in this way for a time it is important to pass on and not merely take counting by adding one thing to another....you should now teach the child as follows: "This is something which is ONE. Now you divide it like this, and you have something which is TWO. It is not two ONE's put together but the two come out of the ONE." And so one with three and four. Thus you can awaken the thought that the ONE is really the comprehensive thing that contains within itself the TWO, the THREE, the FOUR, and if you learn to count in the way indicated in the diagram [p. 95], then the child will have concepts that are living. He thereby comes to experience something of what it is to be inwardly permeated with the element of number.

...You should develop the child's thinking by means of external things which he can see, and keep him as far away as possible from abstract ideas.

The children can then gradually learn the numbers up to a certain point, first...up to 20, then up to 100, and so on....I should like to emphasize that this method of counting, real counting, should be presented to the child before he learns to do sums....

Arithmetic too must be approached out of life. The living thing is always a whole and must be presented as a whole first of all....If you are walking towards a distant wood, you first see the wood as a whole, and only when you come near to it do you perceive that it is made up of single trees....You never have in your purse, let us say, 1, 2, 3, 4, 5 coins, but you have a heap of coins....

Thus in your teaching you must not start with the single addenda, but start with the sum, which is the whole, and divide it up into the single addenda. Then you can go on to show that it can be divided up differently with different addenda, but the whole always remains the same....

Starting with the minuend and the subtrahend and working out the remainder is a dead process. But if you start with the minuend and the remainder and have to find the subtrahend, you will be doing subtraction in a living way....

Multiplication: "Here we have the whole, the product. How can we find out how many times something is contained in this product?" This thought has life in it....We must never, never separate thinking from visual experience, from what the child can see, for otherwise we shall bring intellectualism and abstractions to the child in early life and thereby ruin his whole being....

You should teach the children to count from their own bodies as I have described...first with the fingers and then with the toes...it would be good to accustom the children actually to count up to 20 with their fingers and toes, not on a bead-frame....this is then a meditation, a healthy kind of meditating on one's own body.

Page 143: If a whole is divided in a certain way, what is the amount of the part? And you have only another conception of the same thing as in the question: By what must a number be multiplied in order to get a certain other number?

Thus, if our question refers to dividing into parts, we have to do with a division: but if we regard it from the standpoint of "how many times..." then we are dealing with a multiplication. And it is precisely the inner relationship in thought which exists between multiplication and division which here appears most clearly.

But quite early on it should be pointed out tot he child that it is possible to think of division in two ways. One is that which I have just indicated; here we examine how large each part is if we separate a whole into a definite number of parts. Here I proceed from the whole to find the part: that is one kind of division. In the other kind of division I start from the part, and find out how often the part is contained in the whole: then the division is not a separation into parts, but a measurement. the child should be taught this difference between separation into parts and measurement as soon as possible, but without using pedantic terminology. Then division and multiplication will soon cease to be something in the nature of merely formal calculation, as it very often is, and will become connected with life.

So in the first school years it is really only in the method of expression that you can make a difference between multiplication and division; but you must be sure to point out that this difference is fundamentally much smaller than the difference between subtraction and addition. It is very important that the child should learn such things.

At first one should endeavor to keep entirely to the concrete in Arithmetic, and above all avoid abstractions before the child comes to the turning point of the ninth and tenth years. Up to this time keep to the concrete as far as ever possible, by connecting everything directly with life.

Discussions with Teachers

Pages 44-46

Phlegmatic/Addition:
- Children must first be able to count
- Count the stones to get the sum
- Make several little heaps of stones
- $27 = 3 + 7 + 12 + 5$
- With choleric child: reverse process:
- Arrange stones in piles so that $3 + 7 + 12 + 5 = 27$

Melancholic/Subtraction:
- Child is asked to count the stones
- Child counts 8
- "I don't want 8, I only want 3. How many must you take away?"
- Child discovers that they must take away 5
- With sanguine child: reverse process:
- Ask what has been taken away, let child say that $8 - 5 = 3$

Sanguine/Multiplication:
- Start with a number of stones that is divisible by a whole number
- "I have 8 stones. How often can you find 8 stones in the 56?"
- Hence multiplication leads to a dividing up. Child finds answer is 7.
- With melancholic child: reverse process:
- Ask what number is contained 7 times in 56.

Choleric/Division:
- Working back to the dividend, from the smaller number to the greater
- Pile of 8: I want to know in which number you can find 8 seven times.
- With phlegmatic child: reverse process, ordinary division.

"What if my child doesn't seem to be responding?"

Being Your Child's Teacher

This guide is intended both for home-schooling parents and for classroom teachers, so you might be working with one child or with a group. If you are teaching your own child, please bear in mind that you have chosen a challenging, yet joy-filled and rewarding, path. We often hear from home-schooling parents whose children are not responding to them as teachers, thus making the teacher-learner relationship a struggle to maintain. We recommend that you consciously separate yourself as teacher from yourself as parent from the very beginning. But how does one do this?

First of all, create a classroom space before classes even start. This space should be quiet and, as much as possible, without distractions. When you go into this space, you become the teacher by standing in front of the child/ren while they sit at their desk(s) or table(s). This might feel uncomfortable to begin with, but if you maintain this posture, it should soon become second nature to you both.

Secondly, think like a teacher. You might even consider wearing certain clothes to teach in as well as dressing your child in "school clothes." This will make school days different from non-school days.

Next, create a ritual for entering into the day's classroom time. Ritual provides a dependable, predictable framework for each day's work. Having a daily ritual separates school work from the rest of the day; it signals to the child that, "This is when we focus on learning." Morning Exercises will start, every day of the year, with a similar set of activities, and will end with the same verses. For example, a Morning Lesson will always end with the "song to end Morning Lesson." Dedication to your child's schooling also means that you will remove and turn off your phone, the television, the radio, Alexa, etc. This signals to the child that you consider this time with them to be important and worthy of your complete attention.

Use an easel [a sturdy music stand will do] to display your sample work. Having a large chalkboard mounted behind you will also reinforce your image as "teacher." If the child needs help, stand beside them and bend down to the desk, but avoid sitting next to the child or putting your arm around them, even while you're telling a story. These are sympathetic gestures that signal "parent," and they will interfere with your efforts to be the teacher.

Of course, you know your child better than anyone else does. You will be able to pay attention to your child's actions and know what is needed. If your child has been working happily for several weeks and suddenly bursts into tears in the middle of a lesson, you'll know it's time to check for a fever. But if your child has a pattern of crying in several lessons, analyze the situation: Is the material too hard? Are your expectations too high? Does your child want your attention [perhaps in an unconscious effort to avoid the work?]

In other words, one of your tasks in home-schooling will be to decide to develop a teaching persona that is different from your parenting persona. In our experience, drawing this line at the very beginning of your home-school efforts will make it much easier to maintain throughout the process. In this way, your child will immediately know that having school at home is different from the at-home activities and relationships they are familiar with. This is not to take away from the joy of learning, but to help you establish rhythm and accountability and the structure needed to impart these enriching and fulfilling Waldorf lessons.

About the Author

VIVIAN JONES-SCHMIDT has been teaching for over thirty years. With a Master's degree in Education, she has spent most of her teaching years at the Charlottesville Waldorf School and is now starting her fifth year at Lotus & Ivy, Waldorf-inspired virtual classes. She has served in many leadership positions, mentored several teachers, and was one of the earliest members of the Editorial Board of Renewal: A Journal for Waldorf Education, for which she has written several articles about Waldorf Education. She has written many poems and songs for her classes, and a few of her middle school plays have been gathered into the book Three Plays for Small Classes.

BRING THE MAGIC OF WALDORF LEARNING HOME

Live, Interactive Virtual Classes taught by Experienced Waldorf Teachers

If you and your child have loved this curriculum, there's so much more to explore with us!

- Real-Time Connection
- Low-Tech, High-Touch Approach
- Rooted in the Seasons and Developmental Stages of Childhood
- Small Class Sizes & Gentle Guidance

Explore our class offerings and join a community that cherishes head, heart, and hands learning, rhythm, and the wonder of childhood.

📞 321-866-6860

🌐 www.lotusandivy.com